...ne World:

...Coverage
...Networks

Television's Window on the World:

International Affairs Coverage on the U.S. Networks

James F. Larson

University of Washington

Ablex Publishing Corporation
Norwood, New Jersey 07648

Library of Congress Cataloging in Publication Data

Larson, James F.
  Television's window on the world.

  (Communication and information science)
  Bibliography: p.
  Includes indexes.
  1. Television broadcasting of news—United States. 2. Foreign news—
United States. I. Title. II. Series.
PN4888.T4L36   1984   070.1′9′0973   84-15859
ISBN 0-89391-142-9
ISBN 0-89391-312-X (pbk.)

Ablex Publishing Corporation
355 Chestnut Street
Norwood, New Jersey 07648

Contents

Acknowledgments   vii

Introduction   ix

Chapter 1: U.S. Television as an International News Medium   **1**

The U.S. Networks and International News   1
International Concerns Affecting U.S. Network Television   14
A Conceptual Approach to the Study of Television News   20
The Importance of the Context and Conceptual Framework
    for This Study   32

Chapter 2: International News Content on Network Television   **34**

Research Methods   34
International Affairs Content on Network Television News   39
Major Story Formats   42
Thematic Content of International News Stories   45
The Rank Ordering of International Stories   46
Packaging International News: A Characteristic News
    Broadcast   49

Chapter 3: The News Geography of Network Television   **51**

Nations and Territories on Network News   53
World News Leaders During 1972-1981   55
Coverage of Major Geographical Regions   61
Geographical Patterns in Network News Coverage   90

**Chapter 4: Coverage of Developed, Developing, and
Socialist Nations**   **93**

The Problem of Third World Coverage in Major Western Media   93
Existing Research   95

The Categorization of Nations for This Comparison 97
Coverage of Developed, Developing, and Socialist Nations 97
Network Television Coverage of Three Worlds 109

Chapter 5: Some Influences on Network World News Coverage 113

Conceptual Approach 113
Methodology 117
Findings 119
Summary 128

Chapter 6: Television News and the Foreign Policy Process 129

Network Television as an Observer of Foreign Policy News 130
Network Television as a Participant in the Foreign
    Policy Process 134
Television News as a Catalyst in Foreign Policy 140

Chapter 7: What Sort of Window on the World? 144

Summary of Major Findings 144
The Implications of the Findings 149

References 152
Additional References 159

Appendix A: Number of Newscasts and Stories Sampled, by Year 165

Appendix B: Content Analysis Units and Coding 166

Appendix C: Reliability of the Television News Index and Abstracts as a Measure of Coverage of Nations, 1972-1976 170

Appendix D: Nations and Territories of the World 171

Appendix E: Bivariate Regression Analyses with Time as the Independent Variable and the Three Major Story Formats as Dependent Variables, All Networks, 1972-1981 179

Appendix F: Fifty Most Frequently Mentioned Nations, by Network (ABC, CBS, NBC) and by Year (1972-1981) 181

Author Index 189

Subject Index 191

Acknowledgments

This book is the culmination of more than 7 years of research on network television's coverage of international affairs. During those years a number of individuals and organizations contributed in one way or another to the ongoing effort.

Professor Everett Rogers of Stanford University initially suggested publishing the research findings in book form. He also guided the conceptualization and design of the early research, along with four other Stanford faculty members, Professors Richard Brody, William Paisley, Edwin Parker, and William Rivers. Andrew Hardy, then a fellow graduate student, helped design and conduct the original content analysis of television news. Professors Robert Hornik, John Mayo, and Lyle Nelson all provided encouragement or some form of tangible support when it was needed in the early stages of the research.

Others made it possible to continue and significantly expand the scope of my research after arriving at the University of Texas in 1979. For the past several years I have collaborated on television news research with Professor Emile McAnany, a colleague in the College of Communication. His scholarly interest in the topic predates my own and the research described in this book benefited immeasurably from our work together. His comments and suggestions concerning the final draft manuscript were especially helpful.

In early 1980, a grant from the Television and Politics Study Program at George Washington University provided financial support for a Research Assistant and for greater utilization of the Vanderbilt Television News Archive. Professor William Adams, director of the George Washington University program, invited me to contribute a chapter to one of his two edited volumes on television's international news coverage. As shown by his own work, Professor Adams is an extremely capable and energetic scholar. His advocacy and support of better television news research has enriched my own inquiry along with that of many others.

At about the same time, Professor George Gerbner, dean of the University of Pennsylvania's Annenberg School of Communications, invited me to present some of my research findings at an international conference, *World Communications: Decisions for the Eighties,* held in Philadelphia during May of 1980. The interaction with scholars, government officials and media professionals at that conference provided both encouragement and a better sense of direction for my research.

The University of Texas at Austin, through its University Research Institute, also provided essential support for the completion of this book. In August of 1981 I received a Special Research Grant to hire a graduate research assistant. That person, Douglas Storey, became intimately familiar with content coding, but far more importantly he became an active and valued colleague in the research. More recently, the University Research Institute provided a Summer Research Award which allowed me to complete writing this book during the summer of 1983.

During the course of my work on the book, I consulted Professor Melvin Voigt, editor of the Ablex Communication and Information Science Series, on several occasions. He provided help in a variety of ways, including a careful review of the completed manuscript.

I also want to thank my parents, who early in life instilled in me the value of perseverance in completion of a job. The writing of this book required more than a little persistence. In at least this important respect, it was no different from some of my regular childhood tasks around home, however tedious or mundane I considered them at the time.

My wife and daughters, Leone, Grete, and Katie, all contributed in a unique way to this book. My thanks to each of them for their patience and love as well as their occasional impatience and often welcome distraction from the research and writing.

Finally, I dedicate this book to the vast majority of the world's population in the developing nations of Asia, Africa and Latin America. Television must become a much better window to adequately convey news of their world to viewers in the United States and other developed nations.

Introduction

Television is often described as a "window on the world." The metaphor is particularly appropriate to describe television's coverage of international news, which is the subject of this book. It evokes the visual power of television in comparison with other major news media. It implies the global scope of international affairs. Finally, and not least, it conveys the notion of news as a frame which delineates a particular view of the world.

Although this volume focuses on U.S. television news, the window metaphor might apply equally well to television systems in other nations and cultures of the world. This broader, global relevance is especially important in an era when all major media, including television, are under increased international scrutiny. An important premise of this study is that current international concerns, particularly those voiced by developing nations, form part of the context for understanding U.S. television news, now and in the future.

The purpose of this book is to examine television news as a "window on the world" for the United States during the period from the early 1970s through the early 1980s. That time span encompasses a number of significant changes relating to television's coverage of international affairs. There were improvements in the technology for gathering visual news, and along with them changes in the practices of foreign correspondents. Compared with their predecessors in the early days of television, they became much more mobile, likely to travel from one breaking news story to another with fewer ties to a single country or city as a home base. An increasing proportion of Americans came to rely on television as their principal source of international news. The economic picture also changed, and all of the network news operations became profitable. In the realm of international affairs, there were wars, crises, negotiations and other events that helped to keep international news a priority for the U.S. television networks. Partly as a consequence of these developments, there were important changes in the conduct of international diplomacy and the process of foreign policy formation.

Considering the rapid evolution of the medium in the 1970s, it is hardly surprising that television news is viewed as a principal catalyst in the debate about news in general (Golding and Elliott, 1979, p. 10). Television's visual immediacy, its large audiences, and its political uses have all contributed to the ongoing debate. Internationally, developing nations have voiced concerns about qualitative and quantitative imbalances in the flow of news, attributed in part to the global dominance of four Western news agencies and two newsfilm agencies. These concerns have led to the discussion of such related issues as the training and rights of journalists, the role of new technology for the news media, and the question of alternative definitions of news and news selection criteria.

This book examines network television news by focusing on three broad areas. The first and principal focus of the book will be on the content of international affairs coverage provided by the ABC, CBS and NBC television networks on weeknight news broadcasts during the decade from 1972 through 1981. As others have noted (Paletz and Entman, 1981), content is crucial to an understanding of the political role of media. Chapters 2, 3 and 4 all describe various dimensions of international news content, both quantitative and qualitative.

The second focus of the book is on influences that shape the reality of the world as framed by television's window. Chapter 5 examines in particular the influence of satellite communication, and the global deployment of network bureaus and correspondents for major news organizations, on televised coverage of international news.

Finally, a third focus of the book is on the consequences of television's coverage of international news, particularly as they relate to the foreign policy process. Chapter 6 describes the changing role of network television in relation to public diplomacy and foreign policy. Chapter 7 provides a short summary of the major findings reported in this book and comments on their significance.

The purpose of the initial chapter is twofold. First, it sets forth a more detailed context for the study by describing some important aspects of the growth of television as an international news medium. This background includes not only changes in network news within the United States, some of which would be apparent to regular viewers, but also the growth of international concerns which led to the current "crisis" in international news (Richstad and Anderson, 1981). The second purpose of the introductory chapter is to outline the conceptual approach which guides the empirical research reported in later chapters. Drawing on previous work by sociologists, political scientists and communication researchers, this final portion of the chapter delineates a conceptual framework for better understanding the role of television as an international news medium.

Chapter 1

U.S. Television as an International News Medium

Television serves as a "window on the world" not only for viewers in the United States, but also for an increasing number of people around the world. The emergence of television as an important international news medium parallels the general growth of television in different parts of the world. A good indicator of the general trend is the continuing expansion of Visnews and UPITN, the two leading distributors of newsfilm in the world. Visnews alone puts more than 400 cameramen in the field daily and distributes stories on film and videotape to 200 broadcasters in almost every nation where television exists. Visnews maintains 16 bureaus and has reciprocal arrangements with television operations all around the world. UPITN is the major global competitor with Visnews. The trend in both organizations is toward more use of videotape and the use of satellites to transmit news around the world (King, 1981).

Although the rate of change varies from region to region and undoubtedly is faster in the more economically developed nations and regions, the development of television as an international news medium is unquestionably a process of global proportions. Therefore, the context for this study of U.S. network television must include relevant international developments as well as those changes which have occurred within the United States. The following pages examine, first, the major factors behind the emergence of television as a leading source of international news in the United States, and second, those international concerns with direct ramifications for the future performance of U.S. television as an international news medium.

THE U.S. NETWORKS AND INTERNATIONAL NEWS

Over the past decade or more, the three major U.S. television networks, ABC, CBS and NBC have expanded their role as international news media in a number of ways. This point could hardly be lost on regular viewers of network television, because all three networks have spared little effort to portray themselves to the public as important sources of international news. In the terminology of marketing and advertising, they are attempting to position themselves in the consumer's mind as leading sources of international news. All three of the early evening network news broadcasts stress the global or worldwide nature of their news coverage in their title or graphics. ABC titles its half-hour program "World News Tonight," while both "The CBS Evening News" and "NBC Nightly News" use a globe or map of the world in the visual graphics introducing and closing the programs. As of January, 1983, ABC was promoting its television news programs with the suggestion that ABC News is "Uniquely Qualified to Bring You the World." It is probably no exaggeration to say that images of a globe or world maps have been the primary graphic device used to position and promote the early evening network news over the past decade.

In an important sense, one purpose of this book is to examine how well the three networks have lived up to the claims of their own promotional efforts. How completely and how well have they conveyed information about international affairs to the American public? Clearly, there have been changes in network television's gathering and dissemination of international news. They include technological developments, changes in the economics of commercial network news, the growing political role of network news in relation to foreign policy, and shifts in the media usage habits of the public. The following pages explore each of these changes in turn.

Technological Developments

Communication satellites are the most dramatic example of a technological influence on the coverage of international news by the U.S. networks. However, another closely related influence was the development of miniaturized electronic newsgathering equipment. These portable cameras and videotape editing equipment greatly simplified the process of video transmission from overseas locations.

Communication Satellites

During the 1970s, the Intelsat global satellite system expanded from a fledgling system to one that encompasses a majority of nations in the world. At the start of the decade, only 24 nations possessed earth stations and a few more had access to the system through terrestrial connections with countries having earth stations. By the end of the decade, more than 135 nations were using Intelsat system services full time (Comsat, 1980, p. 33). Over the same time period, the cost of television news transmission via satellite decreased dramatically, minimizing cost as a factor in network decisions concerning the transmission of visual news. Mosettig and Griggs[1] report that "transmission costs from such points as Tehran, Cairo, Tel Aviv, and Rome have dropped by almost half compared with a decade ago. The drop has been even sharper from Europe" (Mosettig and Griggs, 1980, p. 74). They are referring, of course, to the cost of transmitting television news items from different parts of the world to the United States, a principal news hub of the world. Since transmission costs relate to demand and volume of traffic, such cost reductions are as yet small comfort to television news organizations in some developing nations which are faced with much higher tariffs.

In principle, the great expansion of Intelsat over the past decade makes it possible for the U.S. television networks to offer timely, if not instantaneous, visual news from a wider variety of locations around the world. Their performance on this score will be examined more carefully in Chapter 5 of this book. Here, some historical perspective on the use of satellite technology by television news organizations illustrates the complexity of the issue.

During the 1960s, government officials, scholars and communication practitioners all expressed optimism regarding the applications of satellite technology by the news media. A group of governmental experts on space communication met in Paris under UNESCO auspices to consider applications of satellite communication. The group agreed on several objectives for the future development of communication satellites, including "Making the flow of visual news in the world more balanced, particularly with regard to providing news coverage to and from, as well as between, developing areas" (UNESCO, 1970). At about the same time, Wilbur Schramm, a well-known communication researcher and UNESCO consultant, found reason to believe "that satellites can potentially make a real difference in news availability throughout the world" (Schramm, 1968, p. 14).

The U.S. network news organizations themselves shared this generally optimistic outlook of the late 1960s. At ABC News, Donald Coe, whose duties included arranging for the ABC News use of all the then-existing satellites, from Telstar to Early Bird, made the following statement:

> We in the news media, particularly television news, naturally want to make the fullest use of every new tool that comes along. Sometimes television is accused of displaying technical advances unnecessarily, using the tools to say nothing. I do not think that charge will be made of our use of satellites. (*Communications Satellites*, 1967, p. 59)

---

[1] At the time their article was published in the Spring 1980 issue of *Foreign Policy*, Michael Mosettig was an associate at the Columbia School of Journalism and a former NBC News producer. Henry Griggs, Jr. was the foreign news producer of NBC "Nightly News."

Has the optimism of the 1960s been justified by the U.S. networks use of satellite technology in the 1970s? How does the U.S. experience compare with the pattern described by Hulten (1973), who studied the early years of the Intelsat system? He concluded that satellites had simply increased the news flow between news centers of the world and had not appreciably influenced flows from peripheral areas to these centers or flows between minor areas. While satellites offer the potential of news from previously inaccessible countries, his study concluded that the new technology does not appear to alter underlying news interests and news evaluation patterns.

In addition to news interests and selection criteria, the realities of international politics also exert some influence on television's use of satellite technology. A network news executive testifying before a subcommittee of the U.S. Senate's Committee on Foreign Relations offered the following glimpse of the problems faced by television news teams overseas:

> Many countries, particularly the third world countries, demand a list of topic areas or subject matters to be covered before issuing visas to a television news team. Then if a subject on the list appears to present even the slightest prospect for embarrassing the government in power, the visas never arrive. They are never overtly refused, they just never arrive at the Embassy where the application was made.

> In other instances, we might find that the local television facilities, almost always government-dominated and usually government-owned, are faulty or unavailable for satellitting [sic] if the story doesn't please the government. For example, satellitting is fairly routine out of Bangkok unless the story involved meets government disapproval; then the facilities inevitably break down just prior to the satellite deadline. (Sheehan, 1977, p. S9806)

Another more recent example of the political difficulties encountered by television news teams occurred during the 1979 hostage crisis in Iran. The Iranian government at first decided to allow news coverage by U.S. network television. Later in the crisis, first CBS, then the other two networks, were denied access to the satellite facilities for broadcasting reports that offended the government of Iran (Quint, 1980).

At a later date, during the Iran–Iraq war, the Iraqi government refused to allow television journalists to use Iraq's satellite facilities. "Videotape had to be driven for 16 hours across the desert to Amman, which meant that the pictures seen on American TV were at least a day old" (Townley, 1981).

Although political influences and news practices may temper the impact of satellite technology on the network news organization, there can be little question that the new technology has had some influence on the international affairs coverage of network television. Several examples illustrate the point.

First, by increasing the availability and lowering the cost of telephone communication with many parts of the world, communication satellites have facilitated the interactive process by which the networks plan overseas correspondents' reports for their weeknight news broadcasts. It is important to remember that network television news is centrally assigned by assignment editors in New York or sometimes other locations (Epstein, 1974, p. 135). Early each weekday, the executive producer puts together a "lineup" indicating what stories are scheduled for broadcast that evening. The lineup, which may change during the day, serves as a guide for preparation of taped video reports, graphics, scripts and narrative. Such changes involve interactive communication between correspondents in the field and editors in New York or another anchor location. Decreasing satellite tariffs and the proliferation of earth stations around the world have made such interaction easier and less costly.

Second, the decreasing cost of satellite transmission is one important factor behind increased network use of their own video reports, decreasing somewhat their reliance on the major news agencies. In this respect it is instructive to compare Epstein's (1974) account of network news operations with more recent descriptions by network news producers (Mosettig and Griggs, 1980; Westin, 1982, p. 74, 75). This trend will be discussed more thoroughly in Chapter 2.

Third, the availability of reliable, broadcast-quality satellite channels made it possible for ABC News to move one of its anchor correspondents to London and on occasion to other locations in Europe or other parts of the world. Whether or not this move actually increased the quality of international news coverage, it illustrates a more general belief in network television news that viewers have come to expect up-to-the-minute reports from a variety of locations around the world as a staple part of the evening network news ritual.

Despite increased and more routine use of satellite channels by network news organizations, the question posed by communication experts in the late 1960's remains relevant in the 1980s. Does satellite technology result in a more balanced flow of visual news around the world? Chapter 5 will explore this question using data on television news content and the growth of Intelsat during the 1972–1981 period.

Electronic Newsgathering Technology

Another important technological development paralleled the growth of the Intelsat satellite system. Thanks to improvements in electronic newsgathering technology, consisting of small, lightweight cameras and videotape editing equipment, all of the U.S. networks made the transition from use of film to the use of videotape. Changing to videotape eliminated the need for film processing and greatly decreased the time required to prepare a visual report and transmit it to New York via satellite. Several years ago, Lester Crystal, then president of NBC News, described the significance of the transition from film to videotape.

> Ten years ago, producers would have been worried about locations from where film of the events could be processed and transmitted. Not today. Water and chemicals won't be needed. Because of the electronic news camera, the lab is obsolete. Even the television station isn't necessary. The closest ground station will do.

> Ten years ago, there were ground stations only in Europe to transmit the daily reports of... a Presidential trip. Today, there are ground stations at every point along the way in Venezuela, Nigeria, India, and Iran, as well as France, Poland, and Belgium (Crystal, 1977, p. 43).

The combination of improvements in satellite and electronic newsgathering technology has been a major factor in changing the role of foreign correspondents for television and the manner in which they gather the news. According to Tom Fenton, a longtime correspondent for CBS News, "The ability that we now have to provide today's foreign news today, with electronic cameras and satellites, means that we quite naturally devote most of our time and effort to the big, breaking stories, to 'hard news'" (Fenton, 1980).

This new role for television foreign correspondents as "firemen" who jet from one breaking news story to another with only loose ties to a home base, is corroborated by individuals who work in New York on the production side of network news (Mosettig and Griggs, 1980). They attribute the change in large part to the two new technologies of satellite communication and improved electronic newsgathering equipment. The question of whether this new style of international news reporting has improved or detracted from network television's coverage of international affairs will be addressed later in this volume, primarily in Chapters 5 and 6.

The Economics of Network Television News

Major shifts in the economics of network television news accompanied the technological changes of the 1970s. As with other programming on network television, most news programs are supported through the sale of commercial time, the value of which depends on the size of the audience attracted to a particular program. The profit picture for network news organizations changed rather dramatically during the decade from 1970 through 1980. In 1970, news programming made no money, but by 1980, ABC broke even and both CBS and NBC showed a net profit on their news operations, after paying out advertising agency commissions and local-station compensation (Pearce, 1980, p. 8).

The early evening news broadcasts of the three major networks are among the most profitable of their news programs. In 1980, an average

30-second commercial on the "CBS Evening News" sold for more than $30,000, an increase from approximately $10,000 in 1970. 10 such 30-second spots are sold during each half-hour news broadcast, for a total gross advertising revenue of $300,000 per program (Pearce, 1980).

In 1979, the "CBS Evening News" made a net profit of $28 million. The net profit equals gross advertising revenue, minus program expenses, advertising agency commissions of 15 percent, and local station compensation for carrying the program. During the same year, net profit for the "NBC Nightly News" was $24.5 million and for "ABC's World News Tonight" was $17.5 million (Pearce, 1980).

One reason for the increasing profitability of network news is that production costs for news programming are low relative to other entertainment programming. This partly explains the decision by all three networks to insert brief news headlines into their prime time programming schedules. These short news segments cost virtually nothing to produce, since the news staffs are on duty anyway, but an average 10-second advertisement in these programs cost over $25,000 in 1980 (Pearce, 1980).

The changing profit picture for news operations also has one major negative consequence for the network news organizations. The profitability of local news operations is the principal reason that local affiliates of each network have refused requests to grant the networks additional time to expand their early evening news broadcasts (Powers, 1981). All three network news divisions would like to expand their weeknight news broadcasts to 45 minutes or an hour, but have thus far been prevented from doing so by the sheer value of the commercial time to local affiliate stations. The rationale most often given by network executives for wanting to expand the early evening news is that world and national events simply cannot be covered adequately in 22 minutes (Epstein, 1974; Friendly, 1982).

Network Television as a Source of International News

During the past two decades, television has emerged as the most credible and widely used source of news in the United States. The trend toward increased public reliance on television for news and public affairs information is clear and unmistakable. However, there has been a debate, both in scholarly circles and in the press, concerning the public's reliance on television versus newspapers. Without question, these two media are the principal news sources for most of the U.S. public, with newsmagazines and radio playing a lesser role.

A significant portion of the existing research on the role of television versus newspapers has been sponsored by either the television or newspaper industries. This sponsorship, along with methodological differences in the conduct of different survey research studies, makes it important to properly qualify statements concerning the credibility, public use, or overall importance of either medium.

The following review of research on the public's view and use of newspapers versus television is selective rather than comprehensive. It is not intended to resolve questions of research methodology. Rather, its purpose is to highlight those research findings that relate most directly to the public's use of television versus newspapers *as a source of international affairs news.*

Television's dominance as a source of international news for the American public can best be explained in terms of five factors:

1. changing trends in the public's use and evaluation of the two media over the past two decades,
2. careful interpretation of public responses to surveys comparing the two media,
3. the relative proportion of international news in television versus newspapers,
4. the relative impact of the two media, and
5. public behavior during periods of "saturation" coverage of international news by network television.

Changes Over Time in Television and Newspaper News

Communication researchers have long known that the introduction of new communication media alters the role of existing media. Bogart (1981) has described many of the changes in newspapers that accompanied the rise of television in the United States.

Despite concerns over deficiencies or differences in research methods (Lichty, 1982; Stevenson and White, 1980) surveys sponsored by both the television and newspaper industries confirm the same basic trend over the past two decades or more. Public viewing of television news has increased while readership of newspapers has declined (Bogart, 1981, pp. 52, 175).

The research most frequently cited to show the increased use and credibility of television is a series of surveys conducted by the Roper Organization for the Television Information Office between 1959 and 1980. The first question in each of these surveys was the following: "First, I'd like to ask you where you usually get most of your news about what's going on in the world today—from the newspapers or radio or television or magazines or talking to people or where?" Multiple responses to the question were permitted. In 1959, 51 percent of American citizens over the age of 18 cited television as a source of most news compared with 64 percent who mentioned television in 1980. Over the same time period, the percentage of people citing newspapers as the source of most news declined from 57 percent to 44 percent.

Another question in the Roper surveys concerns the relative credibility of media and was phrased as follows. "If you got conflicting or different reports of the same news story from radio, television, the magazines and the newspapers, which of the four versions would you be most inclined to believe—the one on radio or television or magazines or newspapers?" In 1959, 32 percent of the respondents said that newspapers were the most believable medum compared with 29 percent who mentioned television. By 1980, 51 percent called television the most believable source, compared with 22 percent who cited newspapers. Surveys conducted by Steiner and Bower corroborate these findings. They show that between 1960 and 1970 television replaced newspapers as the medium which the public feels "gives the fairest, most unbiased news" (Bower, 1973).

Although newspaper industry-sponsored surveys have asked different questions with different results, they tend to confirm the same basic trend shown in the Roper studies. For example, a 1974 study sponsored by the American Newspaper Publishers Association showed readership of daily newspapers remaining nearly constant at approximately 70 percent while public exposure to television newscasts increased from 38 percent of the public in 1957 to 62 percent in 1973 (*News Research Bulletin*, 1974).

Interpretation of Roper Survey Question

Assessment of public reliance on television as a source of international news depends in part on a proper interpretation of the initial question in the series of Roper surveys. Two parts of the question deserve special attention.

First, the term "television news" to which the public responds does not distinguish between network and local news broadcasts. Therefore, the reported findings encompass news at both levels. This aspect of the research is especially important in the case of international news content for two reasons. One is that a higher percentage of the public views local news than the proportion watching network broadcasts. The other is that most local news is broadcast by network affiliates, most of which rely heavily on rebroadcast of international news stories already aired by the respective networks. Epstein characterized the situation in the early 1970s, as follows:

> with the exception of the few unaffiliated stations which obtain their news footage from UPI syndicated news and other independent suppliers, and the noncommercial stations, virtually all the pictures of national and world news seen on television are the product of the three network news organizations. (Epstein, 1974, p. 4)

Second, the phrase "news about what's going on in the world today" may be interpreted by some respondents as implying news of national and international, rather than local, affairs. There is an impressive body of research findings to suggest that this is the case. A review of research on newspapers and television news sponsored by the American Newspaper Publishers Association notes that in the late 1950s, most people in the U.S. perceived that they received most of their general news from newspapers rather than television. However, by the early 1970s that perception had changed. Most people felt they received most of their general news from television, *especially impressions of international affairs* (Weaver and Buddenbaum, 1979). This finding is supported also by a 1966 study in which the public was presented with summary statements concerning various news stories and asked what was the "best way" to find out about it (Bogart, 1981). Television received its highest proportion of responses for items dealing with international and national politics, while newspapers tended to score higher for domestic and local news.

International News Available Through Other Media

As already noted above, local television in recent years has provided little international news coverage except that which is relayed from the national networks or nonvisual coverage received from the major news agencies. This pattern, coupled with the limited coverage of international news by local newspapers in the United States, adds considerable weight to the argument that Americans depend heavily on national television for international news. Bogart (1981) reports results of two studies of a large cross-section of the press, conducted in 1971 and 1977. The average proportion of editorial (nonadvertising) content devoted to international affairs in 1971 was 10.2 percent, and was no doubt inflated somewhat because of heavy U.S. involvement in the Vietnam war at that time. By 1977, the proportion of international news had decreased to 6.2 percent (Bogart, 1981, p. 158). It should be noted that just as studies of television news have shown varying degrees of audience attentiveness during a broadcast, studies of newspaper readership show great variation in attention to different types of editorial content.

News magazines and radio are two additional sources of foreign news. However, both of these media are relatively much less dominant than television and newspapers.

Finally, during the decade covered by the present study, there were important changes taking place in television itself. With the advent of alternative sources of international news on television, such as Cable News network (CNN), the monopoly of the networks on visual coverage of international affairs may change. However, Westin (1982), commenting on the potential threat from cable, pay television and other alternative sources of programming makes a point that others have stressed. He suggests that, if anything, newscasts will grow more important in network schedules because they represent the only type of programming that cannot be taped on a home videocassette for delayed viewing. News, by its very definition, has time value, such that media descriptions of events often lose their news value after a matter of hours. Also, continuous news on television, as with Cable News Network, may not be nearly as effective as continuous radio news because of the requirement that viewers be attending to a television set.

Another potential threat to the network news monopoly lies in the independent gathering of international news by local television stations (Townley, 1982). At the time of this writing it is perhaps too early to judge the significance of that trend. However, given the financial commitment necessary to mount a global newsgathering operation, it would appear that there are numerous practical and logistical questions regarding local news coverage overseas.

Whether such developments as cable, pay television, and local station coverage of international affairs will offer the public more broadly based and extensive coverage of international affairs is still an open question. However, for the time period covered by this study, network television was unmistakably the dominant source of international news for the U.S. public.

The Relative Impact of Television Versus Newspapers

Part of the question concerning the role of television versus newspapers in conveying international news to the American public centers on the relative impact of these two media. While it is tempting to apply the common maxim that "a picture is worth a thousand words," it is a difficult matter to compare two such different media. Television news may be likened to a headline service or the front page of a typical newspaper. Although it can convey only a small fraction of the amount of information contained in most newspapers, it has the capacity to transmit live or only slightly delayed visual sights and sounds. As Bogart describes it, "TV provides a sense of reality that is more complete, more intense, and more valid than what can be conveyed through print" (Bogart, 1980, p. 210).

There is some tentative research evidence that television's capacity for transmitting interesting video material may increase recall or audience information gain (Edwardson, Grooms, and Proudlove, 1981; Katz, Adoni, and Parness, 1977; Stauffer, Frost, and Rybolt, 1981). As yet, however, there has been relatively little research on this question.

In addition to the question of video versus printed information, the relative impact of television versus newspapers depends on specification of different types of effects. For example, the political impact of network television's international news may not depend at all on how the general public uses such coverage. Instead, television's role as a medium for direct public diplomacy and elite dialogue on policy issues may be the principal concern. In terms of its relationship to the foreign policy process, television underwent some dramatic changes during the 1970s. These will be the focus of discussion in Chapter 6.

The Question of "Saturation" Coverage by Network Television

It is likely that the relative importance of television versus newspapers increases for much of the public during periods of "saturation" coverage when all or nearly all of the early evening network broadcasts are devoted to coverage of a single event. In recent years, saturation coverage of international news has occurred in connection with such events as the assassination of Egyptian President Anwar Sadat and the seizure of American hostages in Iran. Bogart (1981, p. 182) questioned the extent to which the public's expressed preference for television in the Roper surveys reflects the psychological impact of major televised events, such as the Middle East war, rather than routine reporting.

In early 1980, during the Iran hostage crisis, the Television Information Office commissioned the Roper Organization to survey the public regarding television's presentation of the Iranian crisis. Instead of asking respondents to indicate where they usually get most of their news about "what's going on in the world today" (Roper, 1981), the survey included the following question: "Where have you been getting most of your news about the crisis regarding the U.S. hostages being held in Iran—from the newspapers, or radio, or television, or magazines, or talking to people, or where?" (Television Information Office, 1980). Seventy-seven percent of the respondents indicated they got most of their information from television, a figure that is 13 percent higher than the 64 percent of Americans who cited television in the regular Roper media survey during November, 1980 (Roper, 1981). On the other hand, only 26 percent of the respondents said they got most of their information about the hostage crisis from newspapers.

Mosettig and Griggs (1980, p. 72) report that audiences for network news programs also increased during the hostage crisis. They report that the combined audience for all three evening network news programs increased from 45 million viewers in October of 1979 to 57 million viewers in mid-December of the same year.

These data clearly show a heightened public reliance on television during the hostage crisis and are consistent with the general notion that television becomes a particularly important source of international news during periods of crisis and the accompanying saturation coverage.

Network Television and the Foreign Policy Process

Part of the explanation for television's emergence as the dominant international news medium in the United States during the 1970s undoubtedly lies in the important events that were covered and the new roles that television came to play in the field of public diplomacy. War coverage, as well as coverage of other crises and more routine international diplomacy changed significantly during the decade of the 1970s.

The decade began with coverage of the war in Indochina. It was the first war that received intensive coverage on television, but in terms of newsgathering technology the coverage was primitive in comparison with present practices. Newsfilm was routinely flown to another Asian country for processing before being sent by satellite or plane to the United States (Mosettig and Griggs, 1980, p. 68). Satellite transmission costs were a factor that influenced the networks to ship a great deal of film by plane (Epstein, 1974, p. 33).

According to accounts by network news producers and correspondents, the nature of television foreign correspondence changed definitively with the 1973 Arab–Israeli war (Fenton, 1980; Mosettig and Griggs, 1980). Although foreign correspondents were still using 16-millimeter color film, Israel had a satellite ground station and there were three transmissions a day during the early days of the war. According to one correspondent who covered the war, it "was the first war to be reported daily on American television screens directly from the war zone" (Fenton, 1980, p. 37).

In addition to coverage of other wars, in such diverse nations as Cyprus, Angola, Rhodesia (now Zimbabwe), Nicaragua, and El Salvador, network television became more actively involved in covering international diplomacy. For example, coverage of President Nixon's trips to China and Egypt was facilitated by newly installed satellite ground stations. Coverage of Henry Kissinger's shuttle diplomacy in the Middle East set a pattern that was to be followed on future diplomatic missions by high government officials. Also, the U.S. television audience saw more frequent interviews with foreign heads of state or senior officials of other governments, as well as coverage of their activities in general.

Egyptian President Sadat's visit to Jerusalem in 1977 marked another important transition in network coverage of international news. By that time, the networks had converted their domestic bureaus from film to portable videotape and were planning to gradually make the same change in their foreign bureaus. Sadat's trip greatly accelerated the transition to videotape overseas (Mosettig and Griggs, 1980, pp. 73, 74).

Iran's seizure of American hostages at the U.S. embassy in Teheran placed network television in an unprecedented and critical role in relation to the formation and conduct of foreign policy. During the hostage crisis there were a large number of "saturation" news broadcasts, during which the networks devoted all or nearly all of their early evening news broadcasts to coverage of the crisis. In addition, the networks periodically preempted other programming at critical periods during the crisis, such as the abortive rescue mission by a U.S. military task force. ABC began regular late evening coverage of the crisis which was subsequently transformed into "Nightline," a permanent half-hour late evening news broadcast.

Some observers have suggested that television was the essence of the hostage crisis in Iran. The government of Iran, the "students" holding the hostages, and the government of the United States all made use of television in various diplomatic moves throughout the long drawn-out crisis (Altheide, 1981).

In summary, the role of network television in relation to the foreign policy process expanded in several ways during the 1970s. Chapter 6 will explore the nature and possible consequences of those changes in more detail.

INTERNATIONAL CONCERNS AFFECTING U.S. NETWORK TELEVISION

A fundamental assumption of this book is that televised coverage of international news in the United States as well as other parts of the world will be shaped in part by issues that are international in scope. More specifically, the continuing international debate over the shape of a "New World Communication Order" highlights a number of issues that are of direct concern to U.S. television and the other major news media in this country. The New World Communication Order is a broad concept, encompassing such issues as new communication technologies, transborder data flow, ownership and control of communication industries, rights and responsibilities of journalists, inequities in the flow of information, and the cultural impact of communication. Those issues and others have been summarized elsewhere (International Commission for the Study of Communication Problems, 1980). The purpose of the following pages is to enumerate and briefly discuss those issues that impinge most directly on the policies and practices of U.S. television news organizations.

The Free Flow of Information Doctrine

One issue of concern to Western news media, including U.S. television, is the internationally accepted philosophy or doctrine underlying the gathering and dissemination of news among nations. Article 19 of the Universal Declaration of Human Rights, adopted by the United Nations General Assembly in 1948, states that

> "Everyone has the right to freedom of opinion and expression; this right includes freedom to hold opinions without interference and to seek, receive and impart information and ideas through any media and regardless of frontiers." (Buergenthal, 1974, p. 74)

In recent years, the principle stated in Article 19 has been commonly referred to as the *free flow of information doctrine*. Substantial disagreement has arisen concerning the scope and possible limitations of the doctrine and it has been subjected to unprecedented international scrutiny.

The crux of the international challenge to the free flow of information doctrine was eloquently stated by Urho Kekkonen, President of the Republic of Finland, at the Tampere Symposium on the International Flow of Television Programs in 1973:

> In the world of communication it can be observed how problems of freedom of speech within one state are identical to those in the world community formed by different states. At an international level are to be found the ideals of free communication and their actual distorted execution for the rich on the one hand and the poor on the other. Globally, the flow of information between states—not least the material pumped out by television—is to a very great extent a one-way, unbalanced traffic, and in no way possesses the depth and range which the principles of freedom of speech require. (Nordenstreng and Varis, 1974, p. 44)

As President Kekkonen indicated, a major problem with the free flow of information doctrine in the international sphere is in its implementation. In practice, he contends, it results in an imbalanced flow of information around the world. Such imbalance or inequity is a major strain in the literature on the international flow of news. It is a finding replicated in virtually every study of international news coverage in major media, but one which did not become a major international political issue until the 1970s.

At meetings of UNESCO and other international conferences during the 1970's, the free flow of information concept was criticized and questioned. Some have suggested that it is no longer acceptable as a conceptual or philosophical basis for international communication (Nordenstreng, 1975; International Commission for the Study of Communication Problems, 1980, pp. 137–145). In its place, proponents of a "New World Communication Order" have advocated a broader concept of a "Right to Communicate" which would stress dialogue and access to communication resources, in addition to the right to receive and impart information.

Waning international support for the free flow of information doctrine is of great concern to U.S. news media, including television. It signals a major shift in the philosophy which has supported international communication activities since World War II. The practical ramifications of such a shift would have direct consequences for the activities of network news organizations.

International News Agencies

The international news agencies have been singled out for special scrutiny in the debate over a "New World Communication Order" (Somavia, 1976; Boyd-Barrett, 1980). Much of the attention has focused on the activities of the four largest agencies: Associated Press, United Press International, Reuters, and Agence-France Presse. Taken together, these four agencies dominate the flow of news information around the world and serve as a primary source of international news for the other media in many nations.

The major international news agencies are relied upon by network television news organizations in two important ways. First, since the networks have only a limited number of permanent bureaus around the world, the news agencies serve as the basic source of initial news information about events around the world. Based on news agency reports and to a lesser extent those of major newspapers, the networks may decide to dispatch correspondents to the scene of breaking or developing stories. Second, the news agencies supply a large proportion of all stories read on camera by anchor correspondents. Based on interviews and observation at the networks, Batscha (1975, p. 122) estimated that 70 to 80 percent of such reports are gleaned from the wire services. Given such heavy reliance on international news agencies, the U.S. television networks have a strong vested interest in their continued strength and viability.

The major international criticisms of the large global news agencies center on the quantity and quality of information available, particularly from developing nations. For example, a Latin American researcher studied the availability of news in several Central and South American nations and concluded that Latin American countries, although they are territorial neighbors, communicate news among themselves according to decisions made by international agencies outside the region (Matta, 1976). Such a pattern is not unique to Latin America. Instead, it illustrates a news dependency likely to be found in most developing regions of the world. The quality of news provided by the major Western news agencies has also come under attack. The criticism of prevailing Western news values applies to all major news media and will be discussed in the following section of this chapter.

One concrete response to criticisms of the dominant international news agencies was the establishment of a Third World News Agency, operating under the auspices of the Yugoslavian News Agency, *Tanjug* (Mankekar, 1981). Whether such efforts will succeed on a major scale and over the long term is still uncertain. As with the existing news agencies, newer organizations must demonstrate their viability in the international marketplace.

As discussed earlier, the pattern of concentration in international distribution of printed news by the major news agencies has carried over into the distribution of newsfilm and videotape. With Visnews and UPITN as the leading suppliers, television news systems throughout the world have demonstrated a strong appetitite for visual coverage of news events (Tunstall, 1977).

Most of the criticisms of international news agencies have also been directed toward the distribution of television newsfilm by Visnews and UPI-TN. In addition, broadcasting unions in several regions of the world have expressed dissatisfaction with the existing arrangements for the exchange of visual news via videocassette or satellite (*Intermedia*, 1981a). For example, the exchange of visual news among nations in the Asian Broadcasting Union has been hampered by high satellite tariffs and the necessary lag time with shipping videocassettes (*Intermedia*, 1981b). Ironically, such dissatisfaction comes at the very time that new technologies such as computers, lasers and satellites are offering the potential to increase the global flow of news and establish new patterns for that flow (Richstad and Anderson, 1981; p. 404). Ultimately, the solution to such problems may well come through greater utilization of the new technologies, as in the successful Eurovision News Exchange.

The news values reflected in the international exchange of visual news have also been the subject of a great deal of criticism. Because television news broadcasts are built around the availability of visual material, television has a stronger tendency than other media to concentrate on violence, crisis or disaster, which often provide visually exciting material.

News Values

As already indicated, the question of news values underlies the international debate over the flow of news through all major media. The question of criteria for the selection of news has focused most sharply in recent years on reporting from the developing nations of the world. Critics of the present system argue that the major media, including television, focus more exclusively than they ought to on wars, other crises and disasters in the Third World. The application of present news values and newsgathering practices results in what is termed the "coups and earthquakes" syndrome (Rosenblum, 1979). In addition, critics have suggested that much coverage from developing nations is oversimplified or cast in a cold war, East–West perspective (Elliott and Golding, 1974).

Communication professionals, scholars and government officials speaking on behalf of developing nations have argued for a different set of values in the selection and dissemination of international news. They suggest that there should be an increase in the proportion of "development news" reported by major media. "Development news" is news of gradual and often long-term social changes which are important to the process of national development. Such changes are often non-spectacular and, very importantly for television journalists, they may be difficult to portray visually. Such a shift in news values would amount to broadening the current Western conception of news to include not only events but also processes. "For instance, hunger is a process while a hunger strike is an event; a flood is an event, a struggle to control floods is a process" (International Commission for the Study of Communication Problems, 1980, p. 157).

The international discussion about news values is a concern for U.S. television journalists for at least two reasons. First, with the advent of new newsgathering technologies and new media, journalists must continually assess the adequacy of their performance, and criteria routinely used for selecting news are an important part of that performance. The need for such assessment is particularly apparent in a period when the available channels for international news and the amount of time available for it are increasing. For the major networks, as well as for such organizations as Cable News Network, the experiences of other nations and a more global perspective may offer an opportunity to broaden and improve the nature of international news coverage in the United States.

Second, shifting international perspectives on news values may be translated into policies that directly affect the newsgathering activities of the U.S. networks and other major media. In other words, nations that favor a different set of news values may also see a greater role for government in relation to the media. Whether through overt censorship or more subtle pressures, governments of some nations may seek to increase their control over the kinds of news that are reported by international media.

Government Control of News Media

The issue of government control of news media and the more general question of the relationship between government and the media have a long history. Probably all governments in the world have at one time or another practiced some form of control. Although this rarely takes the form of direct censorship today, an individual with long professional experience in a major international news agency notes that "All but a few countries in the world make at least some effort to convince correspondents to report things in a favorable light—or to prevent them from reporting anything at all" (Rosenblum, 1979, p. 94).

Although the relationship between government and the media is hardly a new issue, international concern over the question became more pronounced during the 1970s, reflecting the ideological and political distance separating Third World and developed nations. The divergence on basic issues was clearly illustrated at the general conference of UNESCO in Nairobi in 1976. At that conference a draft declaration on fundamental principles governing the use of mass media was put to a vote. The resolution, which received strong support from the Soviet Union, was purportedly aimed at correcting some of the excesses of the international news agencies.

Article 12 of the draft declaration, in particular, became the focus of the intense political debate at Nairobi. It provided that "States are responsible for the activities in the international sphere of all mass media under their jurisdiction" (UNESCO, 1976).

The draft declaration was not adopted at Nairobi. Instead, the issue was postponed and the International Commission for the Study of Communication Problems [2] was appointed to study a broad range of issues relating to international communication. Among other things, the commission studied and rejected proposals for the licensing of journalists. According to the U.S. delegate on the commission, such proposals would strip individual citizens of their right to communicate (Richstad and Anderson, 1981, p. 112).

The Global Impact of U.S. Media

The impact of U.S. media, including television, on the development of media systems around the world is another issue that underscores the international scope of this study of U.S. network television news. The argument for U.S. media influence is articulated clearly by Jeremy Tunstall in his book, *The Media Are American* (Tunstall, 1977). He argues that, because the U.S. has consistently led development of new media, it has had an enormous influence over the subsequent development of the same media in most other countries. For television news, this thesis suggests that basic patterns of news preparation around the world are heavily influenced by the Anglo-American model.

Although evidence concerning the exact nature and degree of U.S. influence on the development of television news in other nations may be mixed, there is a strong and nearly universal perception that American

---

[2]The International Commission for the Study of Communication Problems is often referred to as the "MacBride Commission," after its chairman, Sean MacBride of Ireland.

media have a significant impact abroad (Katz, 1973; Katz and Wedell, 1977; Richstad and Anderson, 1981). This perception partly explains the high level of interest around the world in the performance of U.S. television news, and in the picture it provides of particular nations or regions. Another factor behind such interest is the perceived political impact of network television news in the U.S., a topic that will be explored in Chapter 6.

To the extent that the thesis of U.S. media influence holds true for television news, a careful analysis and better understanding of the U.S. networks may have ramifications for television news in other areas of the world. It appears likely that the U.S. networks will continue to receive some international scrutiny because of their perceived influence. Such attention may in turn stimulate a more thorough and ongoing examination of television news performance by concerned parties within the United States.

A CONCEPTUAL APPROACH TO THE STUDY OF TELEVISION NEWS

The preceding sections of this chapter reviewed developments both within and outside the United States that relate to international affairs coverage by the ABC, CBS and NBC television networks. Such considerations are an important impetus behind this study and help to establish its broad context. However, they do not provide the sort of coherent conceptual framework necessary to better explain television's role as an international news medium. The following pages review some important contributions to theory in the study of international news and describe the conceptual model on which the current research and findings reported in subsequent chapters are based.

Theoretical Approaches to International News

Adams (1978) has identified three areas of inquiry relating to television news. The first is *production research,* which investigates the factors in selection and shaping of news content. The second is *content research,* which looks at various characteristics of the news content disseminated by the media. The third and final category is *effects research,* which investigates the impact of news coverage on its audiences. Although much of this book is devoted to content research, in its conceptual approach there is a broad concern with all three areas. Chapter 5 consists of production research, and Chapter 6, although based partly on content data, discusses the effects of televised international news as they relate to the foreign policy process.

The research literature on international news and the news media more generally, embodies a number of different conceptual approaches. The following treatment of that literature emphasizes two competing models of the news process, one which suggests that news media are engaged in the process of actually creating reality, and the other in which the news process is seen as a selective process that conveys some aspects of "real events."

All three categories of inquiry suggested by Adams are encompassed in one of the most ambitious attempts to develop a general theory of international news. Galtung and Ruge (1965) presented a framework based on a chain of news communication, from world events to personal image. Figure 1.1 relates this news chain to Adams' (1978) categories. The primary focus of Galtung and Ruge's (1965) theory, as with a great deal of other published research, was the first half of the news chain, from events to news media images. In other words, they were more concerned with production and content research than with media effects.

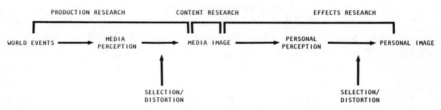

Figure 1.1
The Chain of News Communication

The generality of Galtung and Ruge's theoretical approach should also be underscored. Unlike some other sociologists and communication researchers (Epstein, 1974; Gans, 1979; Schlesinger, 1979), they were not interested in such phenomena as social control in the newsroom, the activities of individual gatekeepers, or the internal processes of news organizations. Instead, their theory treated news media as non-personal and indivisible entities. It also suggested that events have certain qualities or "news factors" that make them more or less likely to become news. These qualities include such characteristics as frequency, lack of ambiguity, meaningfulness, unexpectedness, reference to elite nations, and reference to elite people. Based on a longer list of such news factors, events may be assigned a newsworthiness score. The higher the total score of an event, the higher the probability that it will become news. An event low on one factor would have to be high on another in order to make news. Such a scoring of newsworthiness may then be used to test hypotheses concerning the likelihood that different events will become news.

As Rosengren (1974) has noted, Galtung and Ruge's theory concerns the relationship between *intramedia data* and *extramedia data*. *Intramedia data* are defined as data that can be gathered from the output of news media. Typically, researchers use content analysis to gather such data. *Extramedia data* must be gathered from sources other than news media output and are of two types: (a) those that describe events reported by the news media, and (b) other sorts of data such as trade statistics, population figures, or the number of foreign correspondents stationed in each nation of the world.

The first type of extramedia data would presumably be useful in allowing researchers to compare "what really happened" with news media accounts of happenings. However, a major problem is the scarcity of data that are truly independent of the news media. For certain types of events, such as elections, earthquakes, and railroad or shipping disasters, there are data that appear to be independent of the news media. The limited sources of such data mean that they often will not match the particular questions posed by researchers. Therefore, scholars have tried to obtain or approximate extramedia data in several ways.

Lang and Lang (1971) conducted an innovative study in 1951 by stationing 31 observers along the route of the MacArthur Day Parade in Chicago. The observations of these individuals constituted extramedia data which could then be compared with the televised reporting of the parade.

Another technique for approximating extramedia data is to wait a number of months or years until certain events are universally agreed to have happened. Lippmann and Merz used this approach in a study of the Russian Revolution of 1917–1920. As a standard measurement for the reliability of *New York Times* reports concerning the revolution, they used "a few definite and decisive happenings about which there is no dispute" (Lippmann and Merz, 1920, p. 2).

Rosengren (1970) has suggested that for certain types of events there are registers made up of information compiled from several different news media channels. Although such registers are not independent of the news media they typically would be more complete than any single news channel. Therefore, Rosengren argues that they could be used as a substitute for independent data, "at least when the intramedia data come from only a few channels" (Rosengren, 1970, p. 101).

Rosengren's conceptual approach to the study of news contrasts clearly with that of Galtung and Ruge (1965). While Galtung acknowledges that there are sometimes extramedia data available that can serve as a basis for evaluating image versus reality discrepancies, he maintains that there is generally no such *a priori* baseline to compare with news media coverage, hence news factors are more important than Rosengren's approach indicates (Galtung, 1974, p. 158). In Galtung's view,

> Rosengren's way of thinking leads to a rationalistic model of news. There is a universe, there is selection, the result is an image. In some cases this may be a good conceptualization, but we would in general prefer the opposite way of reasoning: *there is an image, and that image is imposed on reality.* (Galtung, 1974, p. 159)

Galtung's conceptualization of news corresponds closely to the one proposed by Tuchman (1978). She approaches news as a frame, which delineates a particular view of the world. A major theme of her book, *Making News*, is that "the act of making news is the act of constructing reality itself rather than a picture of reality" (Tuchman, 1978, p. 12). The same notion is embraced in the titles of some recent books on television news, including *Creating Reality,* by Altheide (1976), and *Putting 'Reality' Together* by Schlesinger (1979).

The conceptual approach described in following sections of this chapter does not attempt to resolve the tension between these two competing models of the news process, one which stresses the imposition or creation of reality by the media and the other which stresses the role of news media in selecting and conveying certain aspects of real events. Rather, while recognizing that the imposition of an image on reality through news values and ideology is one factor that determines international news coverage, it also acknowledges that economic and political factors play an important role in the creation and dissemination of international news.

A Conceptual Approach for Network Television

The conceptual framework for this current study of U.S. network television borrows from the work of Galtung and Ruge (1965), Rosengren (1970, 1974, 1977a, 1977b) and Tuchman (1978), with certain modifications. It is based on the chain of news communication from events to the ultimate effects of international news coverage. The concepts of flow, process, structure and effects are important to an understanding of the framework as presented in Figure 1.2.

News Flow
News flow is the movement of news along the news chain through any medium. The term connotes a continual and dynamic process, but it should be stressed that most news flow studies, including the one reported here, measure the flow at one point in the news chain. In the past, such studies

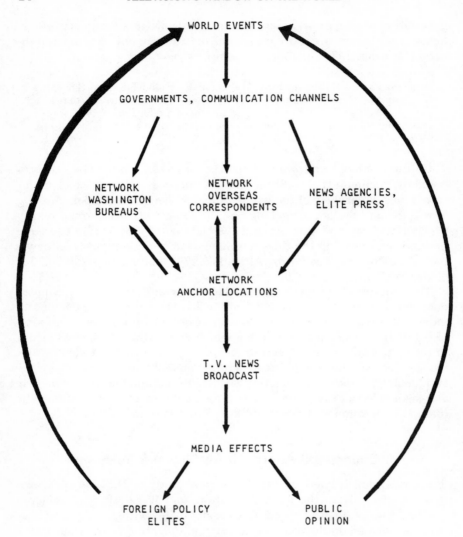

Figure 1.2
News Communication Through Network Television

have often been based on a content analysis of one print medium, typically
a newspaper. Very few studies show how the flow is progressively constructed by gatekeeping at different points in the chain. A notable exception is Cutlip's (1954) study of the flow of Associated Press news from
agency home offices to four Wisconsin nonmetropolitan daily newspapers.

Although the present study measures the flow of television news at one
point in the news chain by analyzing broadcast content, the broader con-

ceptualization of the entire news chain is important to the study and a proper interpretation of its findings. The entire pattern of information flow from world events to network evening news broadcasts forms an important context for interpreting the content of such news broadcasts. Likewise, to use the alternative conceptualization, the overall pattern of activities by the television networks and related news organizations forms the context in which the "reality" created by the news organizations, in this case television news content, can be understood. For example, it makes a great deal of difference whether news is gathered and transmitted to the television networks through a major international news agency, or is gathered and sent to the anchor location by a full-time network correspondent. The latter type of report is considered more visually exciting, more central to the creation and structure of a news broadcast, and more important for maintenance of "audience flow" than a report from a news agency read by the anchor correspondent. Such differences in television news content which reflect more broadly the priorities and processes of newsgathering across the entire news chain will form the basis for analysis of certain findings in later chapters of this book.

A second difficulty with the concept of news flow along a chain of news communication is that it implies a linear, one-way flow of information. Such a conceptualization does not accurately portray the flow of news through television, some portions of which are more interactive in nature. For example, in constructing a network television newscast, personnel from the network anchor location, usually New York, will carry on two-way communication with network bureaus and correspondents around the world. On occasion such two-way communication will enter into the news broadcast itself, in the form of an on-camera dialogue between an anchor correspondent and a foreign correspondent or other individual in another nation. This interactive format was introduced by the "MacNeil–Lehrer Report" on public television and has also been adopted by ABC News for its late evening "Nightline" broadcast. Such an interactive approach was frequently used during early evening network news coverage of the Falkland Islands war between Great Britain and Argentina in the spring of 1982.

A third shortcoming of conceptualizing news flow along a news chain as represented in Figure 1.2 is that it tends to be interpreted as flow along a single chain. In reality, the conception of a news net, extended in time and space (Tuchman, 1978) with multiple chains forming a worldwide network is more appropriate for the study of television's international news coverage. A more complete visual representation of network television's news net would include many chains, connecting New York and other anchor locations with nations around the world. Television's news net is constructed from such elements as satellite earth stations, news agency and network bureaus, and airline connections between nations.

The pattern of the news net varies greatly among the nations and regions of the world. For, example, it is constructed only to catch big stories in regions like Africa or South Asia, but the netting is much finer in Western Europe and the Middle East. All other things equal, more news should originate from regions or nations with a finer spatial news net, a notion that will be examined in more detail in Chapter 5. In view of such considerations, Figure 1.2 should be viewed as only a partial picture of the varied and interwoven structure of the global news net upon which network television relies for its picture of the world.

Finally, a fourth problem with the conceptual approach represented in Figure 1.2 is that it fails to represent the time dimension, which is also an important part of network television's news net. Weeknight and other regularly scheduled network news programs are broadcast on a consistent schedule. One consequence is that certain types of events, referred to by Boorstin (1972) as "pseudo-events", are planned in order to maximize the likelihood that they will receive news coverage. An important aspect of such planning is to time events so that they will mesh appropriately with network newsgathering schedules. A large portion of television's international news coverage consists of events like news conferences, congressional hearings, and Washington, D.C. arrivals or departures of government officials or foreign dignitaries. The White House is the single largest source of such news, but it also emanates from the State Department, Department of Defense, Capitol Hill, and sometimes other branches of government.

News Process
The concept of news process is also central to most studies of the news media. As already mentioned, many studies are entirely devoted to the relationship between the news process and the final news product (Altheide, 1976; Batscha, 1975; Epstein, 1974; Fishman, 1980; Tuchman, 1978). The idea of a news process is important in the study of international news because, if clear boundaries can be set on the process, it becomes possible to distinguish influences that are part of the process from influences that originate outside the news process. Ostgaard (1965) conceptualized the factors influencing the flow of news as either foreign to the news process or inherent in the news process. The former include political and economic factors such as censorship, government news management, ownership of the media, transmission costs for press messages, and advertising. The latter include those factors that are explained by the necessity of making news newsworthy, such as simplification, sensationalism, and audience-orientation (Ostgaard, 1965, p. 40).

The concept of news process refers to all of the activities that take place within the news media, or a particular medium. The news process for each major medium will differ in certain respects. For example writers and editors in radio or television news organizations work under deadline pressure, with broadcasts as often as every hour or half-hour or even continuously, as in all-news radio or television. On the other hand, reporters and editors for newspapers and weekly news magazines contend with an entirely different set of deadlines. The television news process focuses more on sights and sounds than does the process for newspapers. By approaching such cross-media differences in a systematic way, it is possible to describe a different news process for each major medium, including radio, television, newspapers and news magazines.

As shown in Figure 1.2, the network news organizations do not scan "raw" events, as they happen in the real world, except in a limited sense. Instead, they scan the output of three major sources:

1. the *New York Times* and the two major U.S. wire services,
2. their own Washington bureaus, and
3. the stories gathered by their own correspondents and camera crews stationed in selected cities around the world (Epstein, 1974; Batscha, 1975).

Therefore, television news is shaped by a different news process depending on which of the three basic sources is involved.

News Structure

In this volume, when reference is made to the structure of international news, it is not meant to imply that the news itself has a structure independent of the political and economic forces that shape it. It is possible to think of the structure of international news on different conceptual levels. For example, at a very general level as in Galtung and Ruge (1965), structure refers both to news itself and to the political and economic structure of the world in which events become news. In other words, to use the example of the four large transnational news agencies, it is not possible to understand their present structure without addressing the political and economic forces which have historically shaped their evolution. It is no accident that AFP and Reuters maintain a stronger newsgathering presence in Africa than do the two U.S.-based agencies, AP and UPI.

Considered at a different and more specific level, structure may refer to the linguistic or rhetorical structure of individual news stories. For the present study, it is necessary to conceptualize at least three levels of structure relating to international news:

1. the structure of media output or content,
2. the structure of the news media as organizations, and

3. the international and transnational structures that influence the news media.

Structure of Media Output

The output of any given news medium has its own characteristic structure that may be described in varying degrees of detail. The structure of news media content is determined by a variety of influences that are the subject of research on the gathering and production of news. The structure of news content may also have implications for the overall role and influence of a news medium, in this case network television, in the life of a society and nation.

Weaver (1975) provided a thoughtful comparison of the structure of newspaper versus television news in the United States. After noting that both newspapers and television are specialized organizations that cover current events called "news," he identifies three major differences between the two media.

First, he notes that television news is far more coherently organized and tightly unified, both at the level of individual stories and the aggregated level of newspaper editors or television news broadcasts. Partly because television news is organized and presented in time rather than space, as for the newspaper, television stories and entire newscasts tend to present a single, unified interpretation of events. Newspapers, on the other hand, present a great deal more information, both in terms of quantity and diversity.

A second major difference between newspapers and television is the manner in which they narrate accounts of news events. Newspapers use an impersonal narrative voice, which stands in rather stark contrast to the intensely personal narration by television news reporters. Furthermore, the narrative style of television conveys a sense of the reporter's omniscience, encouraging viewers to have an exaggerated view of how much it is actually possible to know and do in the real world.

The third difference between television and newspaper content has to do with the importance television attaches to spectacle. Events that can be spectacularly filmed are more likely to be covered on television than those that cannot (Weaver, 1975). Research by Epstein (1974) and others confirms this tendency.

In this study, as in most other research on international news, structure may also be described in terms of the news story, which forms the basic unit of analysis. The content of both broadcast and print media may be described as a particular arrangement or clustering of stories. A general theory then attempts to explain the relationship between events or what "really happened" and news stories as reported in various media. In other words, clusters of events are compared with clusters of stories.

In the case of international news, which by definition must involve at least two nations, the nation, or sometimes territory, is another basic unit of structure. The substantive interest in most studies of international news is whether certain nations or territories appear in the news, how often, in what context or format, and in connection with what sort of event.

In summary, at an elementary level, the structure of media content may be viewed as clusters of *stories* that describe *events* and mention one or more *nations or territories*. A more elaborate analysis of structure might include all of the actors, individuals and organizations as well as nations, whose activities are reported in international news stories. Such a detailed analysis is presented by Harris (1976b) in his study concerning the "news geography" of the West African wire service of Reuters. In later chapters of this book, much of the analysis of television news content is based on structural considerations like those discussed here.

Structure of the News Process

As already indicated, all major media create news through a process which has been the subject of considerable research over the years. Studies by Tuchman (1978), Fishman (1980), Gans (1979), Golding and Elliott (1979), and Schlesinger (1979) all examine the news process in major media, some more extensively than others. Epstein (1974) was the first researcher to focus on the major U.S. networks, and Batscha (1975) the only researcher to focus exclusively on foreign affairs coverage by the U.S. networks. The collective evidence of such studies is that news process is an important factor shaping content. At the major U.S. television networks, that news process includes such elements as the internal structure of the news divisions, decision-making hierarchies, social control, lines of communication, professional codes, and news selection criteria.

In addition to the scholarly studies of news organizations cited above, there are accounts by media professionals such as the book by Av Westin (1982), an executive producer for ABC News. Such sources provide a wealth of information about the day-to-day operations of the network television news organizations. For example, Westin (1982) describes the interaction between producers and editors, how a lineup of major stories formulated early in the day helps to guide the activities of the entire news staff, and the criteria which determine the selection and positioning of individual stories, whether filed by a correspondent or read by an anchor reporter.

The present research makes no effort to provide a comprehensive description of the television news process. Instead, it will refer to accounts of the news process by both social scientists and media professionals when they bear directly on the question of international afffairs coverage by the networks.

International and Transnational Structures

Conceptually distinct from the structure of the news media are certain international and transnational structures over which the media themselves have very little, if any, control. These are the political and economic structures that tend to define the universe of events that might potentially become news. This universe is not global. Instead, it is limited in practice by such factors as communication channels and correspondent locations, which in turn correlate highly with political, economic and cultural relationships among nations (Harris, 1976b).

Tuchman's (1978) concept of a news net, introduced earlier, illustrates transnational influences on television's international news coverage. The big four transnational news agencies form an important part of the international news net utilized by network television and other media. The location of bureaus and hence the pattern of the news net formed by such agencies is determined by historical, strategic, political and commercial considerations (Boyd-Barrett, 1980). Such factors are international or transnational in character, but they influence network television's news coverage no less than factors which are internal to the networks or subject to their control.

Effects of Television News

A broad conceptual framework for the study of television news would not be complete if it did not somehow account for the impact or effects of news transmission. The effects of mass media have been a major preoccupation of communication scholars, with particular concern in such areas as the effects of televised violence, the effects of advertising to children or other population segments, and the political impact of the media, particularly in election campaigns.

Although the empirical findings reported in the following chapters focus at length on television content and certain major influences on that content, the presumed effects of televised coverage of international affairs are probably the strongest impetus behind the entire study. Television news content and the influences that shape it are only important to the extent that they are assumed to have some important effects.

From the spectrum of possible effects of televised international news, two have been selected for primary consideration in this volume. The first is the effect of such news coverage on public opinion and public knowledge of international affairs, a question of interest to both political scientists and those involved in the field of international education. The second effect is the related question of how television coverage of international affairs affects the foreign policy process and the conduct of modern, public diplomacy. Both of these areas of impact will be examined in more detail in Chapter 6. Here it will be sufficient to note their importance and place in the overall conceptual framework for the study.

International News and Public Opinion According to classical democratic theory, the mass media influence the electorate and through them the officials who act on behalf of government. This notion corresponds closely to the social responsibility theory of the press, a broad treatment of the philosophical and political rationale underlying the system of mass media in the U.S. Two of the six functions attributed to the press under social responsibility theory are:

(1) servicing the political system by providing information, discussion, and debate on public affairs and (2) enlightening the public so as to make it capable of self-government (Siebert, Peterson and Schramm, 1956, p. 74)

Although current conceptual approaches minimize the role of public opinion and the general public in the foreign policy process, (Brody, 1971; Cohen, 1973; Paletz and Entman, 1981) there has been continued scholarly interest and research on the relationship between media content and public opinion. It is generally thought that the quality and quantity of international news available to the public influences at least the potential public role in the foreign policy process (Davison, Shanor, and Yu, 1980).

Adams (1981) compared trends in television news content with changes in American public opinion concerning several countries in the Middle East. He found correlations that "suggested" but did not prove that television and other mass media had caused changes in public opinion. The relative importance of television as a source of international news compared with newspapers and other media, discussed earlier in this chapter, also has a bearing on its relative influence.

Television News and the Foreign Policy Process As noted earlier, the role of television news in relation to public diplomacy and the entire foreign policy process changed in several ways during the 1970s. The changes involved television's roles as an observer, as a participant, and as a catalyst in relation to foreign policy (Cohen, 1963). Perhaps the most dramatic changes during the 1970's have come in television's roles as participant and catalyst in diplomacy. Walter Cronkite has been credited with bringing Egyptian President Sadat and Israeli Prime Minister Begin together or at least facilitating their meeting through a television interview he conducted (Mosettig and Griggs, 1980; Westin, 1982). Leaders of nations in the Middle East and elsewhere have used television as a channel for public diplomacy with increased frequency over the years. In the case of the Iran hostage crisis, there were several incidents in which network television could hardly be described as anything but a direct participant in the foreign policy process.

As illustrated in Figure 1.2, one of the principal effects of television news on the foreign policy process is that it fosters an ongoing public dialogue among foreign policy elites who have direct input into foreign policy decisions (Cohen, 1963; Paletz and Entman, 1981). Chapter 6 will explore the impact of television news on the foreign policy process and will purposely be somewhat more qualitative and speculative in nature than the preceding chapters.

THE IMPORTANCE OF THE CONTEXT AND CONCEPTUAL FRAMEWORK FOR THIS STUDY

The purpose of this chapter was to provide both a broad context and a conceptual framework for the empirical portions of the study which are presented in following chapters. The broad character of both should perhaps be underscored.

The current context for a study of network television news involves interrelated changes in technology, practices for the gathering and dissemination of news, audience uses and expectations of the news, and the relationship of television news to the political policy process. It also includes the twentieth century dominance of Anglo-American news media throughout the world which has generated a high level of concern among developing and nonaligned nations concerning current patterns of hierarchy and dominance in international news. While a study of international news coverage by U.S. network television should obviously be of interest within the United States, it is also unavoidably set in a larger international context.

Given the nature of the context and the scope of network television's activities in gathering and disseminating international news, the conceptual approach needs to be appropriately broad. The approach presented in this chapter is an adaptation of the work of others, principally sociologists, to the study of a particular kind of news content, international news, in a particular medium, television. The purpose of the conceptual framework is to help make better sense out of the empirical findings which follow. As C. Wright Mills has observed, "What you get out of empirical research as such is information, and what you can do with this information depends a great deal upon whether or not in the course of your work you have selected your specific empirical studies as check points of larger constructions" (Mills, 1959, p. 67).

Given the breadth of the context and conceptual approach to the present research, its partial nature should be clear. It focuses primarily on television's international affairs content, as presented in Chapters 2 through 4. While there have been other studies of international affairs content (Adams, 1981, 1982) this is the first quantitative study of U.S. television that covers a full decade. The study focuses to a lesser extent on some influences that shape international news content, as presented in Chapter 5, and in a more speculative vein, some possible political effects of that content in Chapter 6. Without question, the empirical findings which follow are only part, but yet an essential part, of the emergence of television as an increasingly important medium for the international gathering and dissemination of news.

Chapter 2

International News Content
on Network Television

The purpose of this chapter is twofold. First, since the research methods used in gathering and analyzing television news content are important for understanding and interpreting the findings which follow, they are briefly summarized here.[1] The second purpose of this chapter, and its major goal, is to begin a description and analysis of international news on the U.S. television networks, one that will be expanded in the chapters which follow. The quantitative description deals with both the amount and nature of international affairs coverage and is based on a content analysis of weeknight network news broadcasts representing the 10 years from 1972 through 1981. In addition, the study includes some more qualitative analysis of videotaped network news coverage from 1982.

RESEARCH METHODS

As already indicated, the primary research method used to gather information about international news on network television was content analysis. Most of the description contained in this book is based on quantitative measures from the *Television News Index and Abstracts,* published monthly as a guide to the videotape collection in Vanderbilt University's Television News Archive. Some additional data were gathered from a smaller sample of videotaped news broadcasts contained in the Vanderbilt Archive. Although the quantitative data is limited to the 1972–1981 decade, some of the videotapes are more recent.

The Time Dimension

One of the most important methodological features of this study is its longitudinal nature. For practical reasons such as time and limited re-

---

[1] See Appendices A, B, and C for further details concerning the research procedures.

search budgets, much of the research literature on the international flow of news is cross-sectional in nature. In other words, the studies are based on a sample of news content at one point in time, or during a short time interval, rather than over a long time span. For example, one of the early studies of international news content on network television covered a period of 4 weeks during the month of April, 1969 (Almaney, 1970). Most published studies of news media content involve time periods measured in weeks or months, rather than years. This situation leads to potentially incorrect inferences concerning the amount and nature of international news coverage. In addition, it poses the difficulty of comparing the results of a study done at one point in time with those of another conducted at a different time.

Frank (1973, p. 22) called attention to the need for longitudinal studies of international news content on television, in order to avoid some of the distortions inherent in sampling over short periods of time. Also, a longitudinal research design is suggested by what researchers already know about the flow of news around the world. That is, the flow of news in most major media is characterized by relatively stable patterns that are frequently interrupted by such events as war, natural disaster, or political crisis (Schramm, 1964, p. 58). Because this study covers a 10-year period, it allows a differentiation of stable, long-term patterns from brief episodes or disruptions in the pattern of network news coverage. It also establishes a baseline for future research on international affairs coverage by television, and a basis for comparison with international news in other media.

Sampling

The findings presented in this book are based on a random sample of weeknight network news broadcasts during the 1972–1981 period, stratified by year. The total sample includes more than 1,000 newscasts, or approximately 13 percent of all weeknight news broadcasts for each network during the 10 years. This amounts to 7,054 international news stories, more than 2,000 for each of the three major networks.[2]

Sampling error for mentions of nations and territories in the news is acceptably small for the entire sample and for breakdowns of the data by year (Larson, 1978).

Definition of International News

The research literature on international news contains a variety of definitions of "foreign" or "international" news. For example, some researchers define foreign news as only that news reported from outside the United

---

[2] See Appendix A for a complete breakdown of the sample.

States (Hester, 1978). Such a restrictive definition poses two major problems. First, by ignoring the international affairs reporting that takes place in the United States (or in the "home" country of any international news study) it significantly deflates the measure of the overall amount of international news. Second, the definition ignores a central characteristic of international news, namely that it takes place literally "between nations." In reality, only a small part of international news involves just one nation and may be called truly "foreign" news. Even that subset of international affairs reporting is international in the sense that events in one nation are being reported through the media of another nation.

The research reported in the following pages adopts a broad but practical definition of international news. Any news story that mentions a country other than the United States, regardless of its thematic content or dateline, is considered an international news story (Larson, 1978, 1982).

Such a definition, or a close approximation, has been used by other researchers (Stevenson, Cole and Shaw, 1980; Golding and Elliott, 1979) and the conceptualization has several advantages. First, in a content analysis it is easily operationalized through the coding of only "manifest" content (Holsti, 1969) and can be coded with a high degree of reliability. Second, it avoids the problems posed by too narrow or restrictive a definition of international news, as discussed above. Third, since nations are the principal actors in foreign affairs, the coding of all nations mentioned in each news story captures a central dimension of international news.

Although the above definition of international news does not completely solve the vexing methodological problem of how to measure the amount of emphasis given to each nation mentioned in an international news story, it closely reflects the reality that affairs of nations are intertwined in international news. News content therefore becomes one indicator of international interactions.

Reliability

The reliability of data is a concern in two phases of the research reported here. First, since some data were collected directly from the *Television News Index and Abstracts*, rather than from videotaped news broadcasts, it is important to note that the Abstracts are a highly reliable source of data about international news coverage. Prior to any content coding (Larson and Hardy, 1977) audio tapes of 32 randomly selected newscasts between 1972 and 1975 were rented from the Vanderbilt Television News archive. For each of the newscasts, all mentions of nations, territories, and international organizations were coded separately from both (a) the abstracts and (b) the tapes. The two measures were then correlated. The overall Pearson zero-order correlation coefficient between the two measures, for all mentions of nations and territories was .99. The correlation coefficients for individual nations were also very high, indicating that the *Television News Index and Abstracts* are a very strong measure of international news as it is defined for the present study.[3] Second, intercoder reliability was .89 or higher for all content categories used in the quantitative portion of this study.[4]

Units of Analysis

As in most studies of international news, the individual news story or item is the basic unit of analysis. News stories are then aggregated in different ways to describe the amount and nature of international news coverage. The descriptive analysis involves three levels: story level, newscast level, and nation level.

Story Level

The story level description focuses on the number of international versus domestic news stories and on certain characteristics of the international stories, such as length, story theme, story format, and changes over time. All international news on network television may be placed in one of three major format categories, each of which has important implications concerning the newsgathering and selection process of the network news divisions.

First is the anchor report, read from a studio in New York, Washington, Chicago, London or elsewhere by the anchor correspondent and often accompanied by a still picture, map, diagram or artist's sketch. Since the vast majority of such reports originate from the major international news agencies (Batscha, 1975), they provide a measure of network dependence on those agencies for coverage of international affairs.

A second format is the domestic video report, originating live or taped in Washington, New York, or some other U.S. location. The majority of reports in this category originate from the White House, State Department, Pentagon, Congress or the United Nations (Batscha, 1975). They deal with events such as official visits by foreign dignitaries, congressional hearings, State Department briefings, and White House news conferences. In general, reporting in this format deals with nations and events that directly involve the United States government or have somehow become a public issue, most often in Washington or New York.

The third format is the foreign video report, originating live or taped from outside the United States. Such reports are normally prepared by

---

[3] Appendix C presents the correlation coefficients for those nations which received at least five mentions on the audio tapes.

[4] For a detailed explanation of content categories see Appendix B.

network correspondents and transmitted to New York by satellite. Foreign video reports in several respects are the most important part of international news coverage by the networks. They are visual in nature. They indicate the greatest commitment of money and resources by the networks in covering international news. Furthermore, they represent direct newsgathering by the networks which makes the maximum use of satellite and electronic newsgathering technology.

Newscast Level

The newscast level analysis aggregates international news stories in order to describe the extent and dimensions of international affairs content conveyed during entire newscasts. It allows a description of an "average" news broadcast in terms of such characteristics as the number of international stories, their length, rank, formats, theme and the nations with which they deal. As reported earlier based on 8 years of content data (Larson, 1982), the average composite newscast contained 17 items, of which 6 or 7 dealt with international affairs. Two of these would characteristically be foreign video reports, 2 domestic video reports originating from Washington, and 3 would be read by the anchor correspondent. Furthermore, the typical newscast would be more likely to involve certain regions and nations than others, with the USSR, Great Britain, and nations of the Middle East playing a prominent role.

In addition to allowing a description of the "average composite" newscast, the newscast level description also identifies unusual or less typical broadcasts. These include newscasts that contain little or no international news, along with those that are saturated by coverage of an international crisis or other development. Newscast level description is important because it represents the basic format or package in which news is delivered by the networks and consumed by the viewing public.

Nation Level

The third descriptive level deals with the nations involved in international news. It involves the analysis of news geography, which concentrates on the "who" and "where" of network television news; that is, the types of actors and the locations in which activities are reported (Harris, 1976). Nations and governments are the most important actors in international affairs and therefore receive most of the attention in the pages that follow. In addition, data from individual nations may be aggregated to analyze patterns of coverage by major geographical regions, as in Chapter 3, or groupings such as developed, developing and socialist nations, as in Chapter 4.

Quantification of Findings

A final aspect of research procedure important to an understanding and analysis of the findings concerns quantification of the content data. Although all findings of the study are ultimately quantified in terms of news stories as the basic unit of analysis, there are two different approaches to enumeration.

The first approach quantifies results in terms of the number of news stories in which individual nations or aggregated groups of nations are mentioned. Since more than one nation may be mentioned in a given news story this approach leads to percentage tables that may sum to more than 100 percent. An alternative, used on occasion, is to quantify coverage of nations or groups of nations in terms of total national references in the news, in which case the percentages do sum to 100 percent.

The second approach quantifies news stories in terms of their national origin, rather than the nations which may be mentioned in the story. Such an approach is possible only for foreign video reports, but provides an extremely valuable indicator of the allocation of network newsgathering resources.

The remaining parts of this chapter describe 10 years of international affairs coverage on network television, from 1972 through 1981. They include story and newscast level description, leaving the analysis of news geography to Chapter 4.

INTERNATIONAL AFFAIRS CONTENT ON NETWORK TELEVISION NEWS

This chapter deals with the amount and nature of international affairs content on U.S. network television during the 1972–1981 decade. The analysis includes such dimensions as major story formats, thematic content, and story rank. When possible, it traces changes in the extent and dimensions of international news coverage during the 10 years from 1972 through 1981.

Most findings are based on the entire sample of over 1,000 weeknight news broadcasts. However, for certain dimensions of content, data were gathered for more limited time periods or for one or two of the three networks.

It is also important to underscore the global nature of the description. The analysis is presented in terms of total international news coverage, involving all nations and territories in the world. As such, it provides an important baseline for later comparisons with coverage of individual nations, groups of nations, or geographical regions.

Amount of International News

The findings of this 10-year survey confirm that the U.S. networks do devote a substantial amount of attention to international affairs, given the limited time available on their early evening news broadcasts (Adams, 1982). The amount of emphasis placed on international versus domestic news can be measured both in terms of story units devoted to each and in terms of the time allocated to each category of news. Each of these measures of emphasis will be discussed briefly.

Number of International Stories

On the average, weeknight network news broadcasts during 1972-1981 contained a total of 17 news stories, dealing with a range of domestic and international events. On the sampled weeknight broadcasts, the total number of news items ranged from a low of 9 items to a high of 26 news stories.

Of the 17 stories on an average weeknight news broadcast, approximately 7 stories, comprising nearly 40 percent of the evening's news dealt with international affairs. On the sampled newscasts the number of international stories ranged from 1 to 15, reflecting considerable variation in the proportion of any single newscast accounted for by international news.

Broadcasts in which all or nearly all of the news stories deal with international events usually show the power of certain major occurrences to saturate the available news time (Frank, 1973). During the decade under study, the final days of U.S. withdrawal from Vietnam, the assassination of Egyptian President Sadat, the death of Soviet leader Leonid Brezhnev, and the Iranian hostage crisis all drew such saturation coverage by the U.S. television networks on certain days. Other researchers (Frank, 1973) have suggested that saturation events deserve special study.

As shown in Figure 2.1, changes occurred in the amount of international news coverage over the 10 years sampled. The height of each bar in the figure represents the proportion of all network news each year that dealt with international affairs. Each bar is shaded to indicate the composition of the international news in terms of the three story formats used by the networks. The number of stories devoted to international news was high in 1972, averaging more than seven stories per newscast, then decreased to a low of approximately five stories per broadcast in 1974, after which it increased for the remainder of the decade, with some fluctuations. With the exception of 1974, international news accounted for between 34 and 45 percent of all news broadcast by the networks during each year of the decade. Years which contained relatively high levels of international news coverage can all be explained in terms of major international events of concern to the United States, such as the Vietnam war early in the decade,

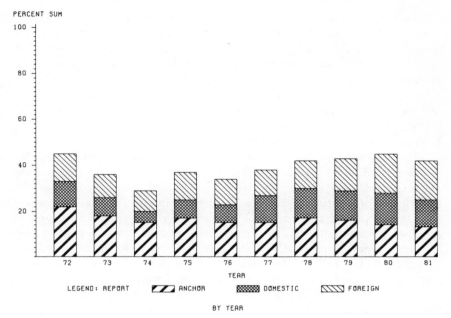

Figure 2.1
Proportion of International News and its Format

coverage of the Middle East in 1979, and attention to Iran in 1980 and 1981. The relatively low attention to international news in 1974 presumably reflected network preoccupation with the Watergate affair and domestic politics. During that year, anchor reports and foreign video reports remained at approximately the same levels as previous years, while domestic video reports showed a sharp drop.

Time Devoted to International News

The amount of time devoted to international news is an alternative measure of emphasis that corresponds very closely to a count of the number of international stories. Although data concerning story length were gathered only for the 6 years from 1976 through 1981, the findings can be extrapolated with confidence to the first 4 years of the 10-year period.

The mean length of an international news story on network television was 1 minute, 28 seconds. However, the mean length varied as follows according to story format.

- *Anchor Report:* 31 seconds.
- *Domestic Video Report:* 2 minutes, 5 seconds.
- *Foreign Video Report:* 1 minute, 57 seconds.

The preceding averages are based on combined data for all three networks during 1976–1981. There is very little difference from one network to another in the duration of international news items of each major format. Nor are there any major changes over time. The one exception to this pattern is that foreign video reports on "ABC World News Tonight" became noticeably shorter during 1980 and 1981, averaging under one and one-half minutes during those years. This change corresponds with a format change by ABC News, which had established Peter Jennings as the London-based anchor correspondent in 1978.

Allowing for station breaks and commercials, each network has between 22 and 23 minutes of air time for presentation of news on the early evening broadcast. During the 1976–1981 period, the networks devoted an average of 9 minutes, 50 seconds of that time to international news. This means that approximately 45 percent of the available air time was spent on international news between 1972 and 1981 while a lower percentage, under 40 percent, of all stories dealt with international news. The discrepancy reflects the tendency of the networks to cover international affairs more often with domestic and foreign video reports than with the much shorter anchor reports, especially during the last half of the decade under study. Table 2.1 shows the amount of time devoted to international news during the weeknight news broadcasts of each network.

Table 2.1
Mean Length of Time Devoted to International News Per Weeknight Newscast, 1976-1981 (in minutes and seconds)

Year	ABC	CBS	NBC	All Networks
1976	7:20	8:22	6:37	7:26
1977	8:56	10:03	7:17	8:45
1978	9:39	11:12	9:42	10:11
1979	10:27	11:27	11:21	11:05
1980	10:33	10:22	10:41	10:31
1981	10:22	10:43	10:27	10:31
1976–1981	9:41	10:29	9:26	9:52
N = (newscasts)	216	216	214	646

MAJOR STORY FORMATS

As indicated above, international news on network television is presented in one of three major story formats: anchor reports, domestic video reports, and foreign video reports. These formats are important because of the close relationship they bear to the newsgathering process through which the networks select and shape the picture of the world presented on weeknight news broadcasts.

During the 1972–1981 period, 42 percent of all international news stories on weeknight network newscasts came in the form of anchor reports, 26 percent were domestic video reports, and 32 percent were foreign video reports. A chi-square test on a 3-by-3 table cross-tabulating story format by network showed that there are no statistically significant differences across the three networks, ABC, CBS and NBC, in the proportion of international news reported in each major format.

The above figures provide a summary description of network television's reporting formats for the entire 1972–1981 decade. However, they do not reveal what changes may have occurred over the 10-year period. Figure 2.2 shows that there were indeed some changes in the relative use of the three major reporting formats over the 10 years.

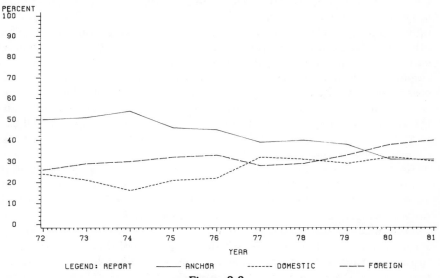

Figure 2.2
Use of Major Story Formats by Year

During the first half of the 1972–1981 decade, the networks used more wire service material in covering international news than they did during the last five years of the period. From 1972 through 1974, at least half of all foreign news stories consisted of anchor reports, but this proportion had decreased to a low of 30 percent by 1981.

By comparison, Figure 2.2 also shows that the networks increased the number of domestic video reports used to convey international news. The proportion of such reports increased primarily during the last half of the decade. The year 1974 was an exception to this trend, with all three networks showing a decrease in the amount of international news conveyed through domestic video reports. The most plausible explanation for this decrease is the heavy emphasis given to coverage of the Watergate scandal during 1974. The networks' Washington correspondents probably devoted less attention to international affairs reporting from the usual Washington sources and more time near Judge Sirica's courtroom. Such a shift illustrates the interdependence of news judgments concerning domestic and international affairs.

Figure 2.2 shows that, as with domestic video reports, the proportion of foreign video reports increased over the decade from 1972 through 1981. Mosettig and Griggs (1980, p. 75) reported a similar trend for the 1976–1979 period, based on analysis of directors' log books at ABC and NBC. This definite trend toward greater use of foreign video reports took place during a decade when, as noted in Chapter 1, there were important technological developments to facilitate the gathering and dissemination of visual news internationally. One development was the improvement in lightweight, portable cameras and videotape editing equipment. The other was the expansion of the Intelsat global satellite system to include earth stations in a much larger number of nations around the world. Corresponding with the growth in Intelsat were dramatic decreases in the cost of satellite transmission of visual news, to the point where satellite costs are now seldom a major factor in decisions about which stories to air (Mosettig and Griggs, 1980, pp. 74, 75). Chapter 5 of this book will explore the relationship of growth in the Intelsat satellite system to network coverage of international news in more detail.

Each of the preceding changes in network news coverage, the increases in domestic and foreign video reports and the corresponding decrease in anchor reports is statistically significant, as indicated by the results of bivariate regression analysis reported in Appendix E.

On the face of it, the changes in relative use of the three major reporting formats show that the networks are devoting more of their own resources to the coverage of international affairs and are relying less on the major news agencies for actual broadcast material. The technological changes mentioned above appear to have facilitated such changes, and intensive coverage of events directly involving the United States, such as Middle East diplomacy and the Iran hostage crisis, may also have contributed to the change. However, it is also well to keep in mind that these changes took place during a period of time which saw increased profitability for network news operations; intense competition among the three networks in hiring anchor and other correspondents; and introduction of format changes and other efforts to garner the largest share of the news audience. Such factors, along with the widely accepted conventional wisdom that visually exciting material and appropriate pacing are necessary to attract and hold an audience, undoubtedly contributed in part to the trends shown in Figures 2.1 and 2.2.

To summarize, the increased use of network correspondents, reporting from Washington, New York, and overseas locations was most likely due to a combination of technological, economic, and purely competitive pressures on the three U.S. television networks. Whether the shift toward greater use of foreign video reports in particular represents a network news commitment toward more broadly based and diverse coverage of international affairs is an open but very important question, to be addressed again in Chapters 4 and 5.

THEMATIC CONTENT OF INTERNATIONAL NEWS STORIES

In compiling the content data file the main theme of each international news story was coded only for the six years from 1976 through 1981. Nine thematic categories were used in coding, and they may be collapsed into crisis and noncrisis themes as follows:

Crisis
1. Unrest and Dissent
2. War, Terrorism, Crime
3. Coups and Assassinations
4. Disasters

Noncrisis
5. Political-military
6. Economics
7. Environment
8. Technology-Science
9. Human Interest

The crisis dimension of thematic content is important for two reasons. First, it has often been observed that television news has a special attraction to crisis or conflict, particularly when it can be presented visually. Such crises as coups, natural disasters and wars often provide the kind of visual material needed by the networks in order to maintain their news audiences. Second, the crisis dimension of television news is important because of widespread claims that coverage of developing nations in the Western media is too heavily oriented toward crisis news. Chapter 4 will explore that question in more depth. The following discussion is intended to establish the overall amount of crisis coverage carried by the networks and some of its basic characteristics.

As shown in Table 2.2., 27 percent of the international news broadcast by the networks during the 6 years from 1976 through 1981 dealt with crisis themes. For CBS News only, thematic data gathered for the entire 1972–1981 decade increased the percentage of crisis news for that network by slightly more than 1 percent. For the 6-year period represented by the data in Table 2.2 there are no statistically significant differences among the three networks in the amount of their attention to crisis news.

Table 2.2
Proportion (percentage) of Crisis Themes by Story Format, All Networks
1976-1981

Story Format	Theme		N = (stories)
	Crisis	Noncrisis	
Anchor Report	28.7	71.3	1,612
Domestic Video Report	14.2	85.8	1,288
Foreign Video Report	36.4	63.6	1,474
All Formats	27.0	73.0	4,374

Table 2.2 also provides corroboration for the notion that commercial television is attracted to crises because of its appetite for dramatic visual material. Of the three major reporting formats, foreign video reports contain the highest proportion of crisis coverage, over 36 percent. Such reports are the most dramatic and visually appealing because the networks have dispatched their own foreign correspondents to report, sometimes directly from the scene of a breaking or ongoing crisis. When it is not possible for the networks to use their own correspondents, they tend to rely on the news agencies for information about crises that may occur overseas. Approximately 29 percent of all anchor reports broadcast by the networks during 1972–1981 dealt with some type of crisis. The lower proportion of crisis themes, 14 percent, in domestic video coverage indicates that routine diplomacy and reaction to events rather than on-the-spot coverage characterizes reporting from Washington, D.C. and other U.S. locations.

Because of the differences in crisis reporting by format, network television news stories about crisis tend, on the average, to be shorter than noncrisis stories. For the last 5 years represented in the content data file, crisis stories averaged 1 minute 20 seconds in length, compared with a mean length of 1 minute 33 seconds for noncrisis news.

THE RANK ORDERING OF INTERNATIONAL STORIES

Television news programs have been likened to a headline service or to the front page of a newspaper. In contrast to newspaper news, television news programs tend to form a unified whole, with a single interpretation of the days events (Weaver, 1975). Within that whole, stories are rank ordered more in terms of how they fit the story line or flow of the program and how that might affect "audience flow" (Epstein, 1974) than in terms of news importance. It should come as little surprise, then, that international news stories sampled over the 1976–1981 period are distributed quite evenly throughout the news broadcasts. Exceptions to this pattern might be the lead story of the broadcast, usually the top news item of the day, and the final story which often contains some element of humor or human interest. Paletz and Pearson (1978) note the tendency of television news stories to be clustered, with stories on related topics appearing in different segments of the news broadcast. This clustering is a strategy consciously promoted by news producers who believe that it helps the viewers to comprehend more of the news. Av Westin (1982), Executive Producer of ABC News gives the following rationale for clustering news stories.

> Stories, including both the anchor's on-camera material and the tape of film pieces from the field, should be combined into a logical progression that threads its way through the day's news. The audience ought to be guided through the news so that it doesn't have to make sharp twists and turns to follow and understand what is going on. My preference is to divide the lineup into segments. In each segment, a narrative of sorts is fashioned, weaving together stories that relate to one another.

> Under this system, it is not unusual to find a story in a newscast's lead segment that the Associated Press places way down on *its* list of importance. (Westin, 1982, p. 66)

The relatively even distribution of international news stories throughout the sampled newscasts holds true for both crisis and noncrisis stories. Based on the original set of nine thematic categories, the only significant difference in mean rank was for human interest stories. They ranked lower, on the average, than stories with other themes, reflecting the well-known tendency to close a network news broadcast with a human interest story. However, an examination of ranks by story formats does reveal a consistent pattern in the structuring of network news broadcasts. To simplify the analysis, stories in the last five years of the sample were divided according to whether they appeared in the first or second half of the news broadcast.

As shown in Table 2.3, all networks place a higher priority on reporting by their own correspondents than on anchor reports, for which they depend heavily on the news agencies. Domestic video reports tend to be used early in the newscast, with an average of 67 percent of such reports appearing during the top half of the program. By comparison, 56 percent of all foreign video reports are aired during the first half of the program. Anchor reports dealing with international affairs are more likely to come later in

Table 2.3
Percentage of International News Appearing in First and Second Half
of Newscast by Major Story Formats, 1977-1981

			Story Format		
Half	Network	Anchor Report	Domestic Video Report	Foreign Video Report	All Types
First Half	ABC	37.6	65.5	52.6	51.1
	CBS	39.1	67.1	60.2	54.6
	NBC	28.6	68.3	56.1	50.3
	All Networks	35.3	67.0	56.0	52.0
Second Half	ABC	62.4	34.5	47.4	48.9
	CBS	60.9	32.9	39.8	45.4
	NBC	71.4	31.7	43.9	49.7
	All Networks	64.7	33.0	44.0	48.0
N = (stories)	ABC	449	368	473	1290
	CBS	466	410	382	1258
	NBC	416	372	410	1198

the program, with about 65 percent of such reports coming in the second half of the newscast.

The data on international news story ranks indicate that, in a typical television news broadcast, the networks will put their own correspondents on the air before using material from other sources. Most often, the leading stories will be filed by one of the U.S.-based correspondents, although foreign correspondents are used nearly as often. Anchor reports, which lack the visual impact of videotape, are inserted lower in the news broadcast.

PACKAGING INTERNATIONAL NEWS: A CHARACTERISTIC NEWS BROADCAST

What does the preceding description of network television's international news coverage mean in terms of an average or typical news broadcast? The question is important because the newscast represents the basic unit or package in which news is delivered to viewers. The audience is not exposed to a steady stream of international news. Nor does it have access to large numbers of international news stories, such as the data file of over 7,000 international news items analyzed here. Therefore, the description of a characteristic news broadcast may help to give some of the quantitative findings a more concrete and understandable form.

On an average network television news broadcast during the 1972–1981 decade, international news accounted for 7 of the 17 news stories broadcast and about 7 minutes, 19 seconds, of the 22 minutes available for news. These figures make it clear that television devotes proportionately much more of its coverage to international affairs than do newspapers (Bogart, 1981, p. 179).

The characteristic newscast of the 1970s would include two domestic video reports and two foreign video reports, more often than not coming during the first half of the news telecast. It would usually include three anchor reports which, more often than not, would be read during the second half of the telecast.

Most often the topics or subjects of network television's international news coverage dealt with war, violent crime, natural disasters, or more routine political and economic interests of the United States around the world. Beyond this general characterization of the news, subjects differed over time. During the first two years of the 1972–1981 decade, half or more of the international stories dealt with the war in Indochina. In 1973, the geographical focus shifted with the outbreak of war in the Middle East. Likewise, late in the decade the early evening news was preoccupied with Iran and the hostage crisis. Overall, the typical network news broadcast was likely to contain news items referring to the USSR, major U.S. allies in Western Europe, Japan, or the Middle East. Much less likely to appear were the nations of Latin America, sub-Saharan Africa, and South Asia.

The concluding story on such a characteristic composite news broadcast was usually of the human interest variety. Britain, other Western European Nations, and Japan appear to be heavy favorites for such closing stories when they come from abroad.

Finally, an important caveat is in order concerning the characteristic newscast and its treatment of international news. It may well be the atypical network treatment of international news rather than the typical approach that illustrates its greatest impact. For example, saturation coverage of international events occurred on a relatively small number of days during the 1970s. Examples include the final day of U.S. withdrawal from Vietnam, coverage of the assassination of Egypt's President Sadat, and several episodes during the Iran hostage crisis including the return of the hostages on the day of President Reagan's inauguration. At such times, international news not only saturates the regular half-hour news broadcasts, but may also pre-empt a great deal of other regular programming. The importance and impact of television news coverage appears to increase greatly during such periods of saturation coverage.

Chapter 3

The News Geography of Network Television

Chapter 1 established that network television news is the principal source of information about international affairs for most Americans. As such, it is an important means by which Americans transform, in Walter Lippmann's words, the "world outside" into the "pictures in our heads" (Lippmann, 1921). This is one reason why a description of television news content is so important. The view provided by this window on the world helps to shape the public's mental map or image of the world.

The question which implicitly underlies this entire chapter is whether the world map provided by network television news corresponds to other maps, including those provided by other media, in this country as well as in other nations. Gerbner and Marvanyi (1977), among others, have shown convincingly that newspapers from different nations have sharply differing patterns of attention to geographical areas of the world. In effect, they convey a different map of the world to their readers. Their study looked at the international news content of 60 daily papers from nine countries of the capitalist, socialist, and third worlds. Among other things they found that the world of British and West German newspapers paid a great deal of attention to Western Europe, Latin America and North America. Eastern European papers, surprisingly, devoted less attention to the Soviet Union than did any other press system, including the United States. The world of the Third World papers was the only one in which the Soviet Union loomed large. The general conclusion from such findings was that readers of all press systems know most about Western Europe, and that all press systems have "blind spots", but in different global patterns. A PBS Television Program, entitled "World: The Clouded Window" produced by WGBH Television in Boston and originally broadcast on February 2, 1978 ("World", 1978), demonstrated that the same phenomenon occurs in television news. Producers of the program obtained film and videotape footage showing how a number of television news systems around the world covered the same event. Despite heavy reliance on the same basic sources for newsfilm and dependence on the same wire services, television reporting differs greatly from one country to another. For example, the PBS program juxtaposed television coverage from The People's Republic of China and The Republic of China (Taiwan) as it was broadcast following the death of Mao Tse-Tung in 1976. The announcer for the People's Republic of China television news read the following script:

> As the entire nation mourns the death of Chairman Mao, the work force and the management of the Peking Petrochemical Company expressed their deepest sympathy for our great and beloved leader; and as a mark of their respect, they resolved to redouble their efforts at work, following the Marxist, Leninist, and Maoist doctrine in the achievement of unparalleled results, as willed by Chairman Mao. They will strive to fulfill Chairman Mao's will—to carry on the class struggle and to hold to the proletarian doctrine as they fight, fight, fight to achieve success.

In contrast, the television reporter for The Republic of China television news gave the following announcement:

> The ambassador to the United States from Taiwan has made a speech in which he referred to the puppet regime in Mainland China, even before the death of Mao Tse Tung. Since his death, the situation has deteriorated, especially in the big cities, right across from north to south.

The PBS Program ("World", 1978) contained other such examples that showed how not only the script, but the editing of newsfilm and other visual material differs from one nation to another in coverage of the same event.

This chapter sketches the world map provided by U.S. network television during the 1972–1981 period by focusing on the news geography of the early weeknight broadcasts. Following Harris (1976b), news geography refers to both the major actors in international news and the locations in which activities are reported. Here, the news geography of U.S. network television is discussed in three ways.

First, since nations are characteristically the most important actors in international news stories, news geography is described through a general discussion of how nations referred to in the news reflect important aspects of such news coverage. The general discussion provides a context for interpretation of more specific findings which follow.

Second, news geography is discussed in terms of those specific nations which received the greatest amount of attention during the 1972–1981 period. This description places particular emphasis on patterns of coverage among "news leaders," those nations which received the most extensive coverage during the decade.

Third and finally, the chapter outlines the geographical patterns of news coverage through analysis of news about major regions of the world.

This concluding section includes both comparisons among the major regions and a description of patterns of coverage within each region.

NATIONS AND TERRITORIES ON NETWORK NEWS

As already indicated, major news media characteristically refer to nations and territories or their leaders in reporting international news. In this chapter and those that follow, all references to countries or nations refer also to dependent territories or possessions, unless otherwise indicated.[1] For example, a mention of Hong Kong or Northern Ireland, neither of which are nation states, is treated the same as a mention of Egypt or the USSR.

Such an approach is more complete than one which is restricted to nations alone and it is necessary because, on occasion, territories may be important sources of international news. An excellent case in point is the Falkland Islands, which became the focus of network television news for an extended period of time during 1982.

As a practical matter, combining data for both nations and territories poses no problem for interpretation of the research findings. Altogether, dependent territories were mentioned in only 3 percent of the news stories sampled for this study, and approximately two-thirds of those mentions involved just two territories, Northern Ireland and Hong Kong. Other dependent territories were involved in only 1.2 percent of the sampled stories.

At a general level the nature of international news coverage on network television can be described along two dimensions. The first is whether or not news stories involve the United States, and the second is the number of nations referred to in the stories.

The United States is referred to in about 60 percent of all international news stories broadcast by each network. Most often, mention of the U.S. indicates some direct American involvement or interest in the affairs reported. For some analytical purposes, it is helpful to ignore mentions of the United States. Table 3.1 shows the number of nations mentioned per international news item, differentiating between references to the United States and foreign nations.

Twenty-three percent, or nearly one quarter of all international news stories broadcast by the networks during the 1972–1981 period mention only a foreign nation and do not explicitly involve the United States. Such news characteristically reflects a news judgement that events occurring in some nation are newsworthy in their own right, apart from any involvement of the U.S. or other nations. If international news is taken to mean

---

[1] For a complete list of nations and dependent territories used in this study, see Appendix D.

Table 3.1
The Number of Nations Referred to in International News Stories,
With and Without Including the United States
All Networks, 1972-1981

Nations Mentioned	N = (nations)	% of All Stories[a]	N = (stories)
1. One Foreign Nation	1	23.0	1591
2. One Nation and U.S.	2	32.8	2271
3. Two Foreign Nations	2	12.4	857
4. Two Foreign Nations and U.S.	3	14.6	1007
5. Three Foreign Nations	3	4.0	280
6. Three Foreign Nations and U.S.	4	5.8	403
7. Four or More Foreign Nations	4 +	2.1	148
8. Four or More Nations and U.S.	5 +	5.2	361
		100.0	6918

[a] Column percentages do not total exactly 100 percent because of rounding error.

Note: Missing cases in this and subsequent tables are accounted for by some stories which mentioned the U.S. and international organizations such as NATO or the United Nations, and by coder error.

literally news "between nations," items that mention only one nation may be considered more genuinely foreign, rather than international in nature.

Another 33 percent of all international news on network television mentions only the U.S. and one other nation. In general, such news is concerned with bilateral affairs between the U.S. and other nations.

As shown in Table 3.1, about 32 percent of all international news stories mention three or more nations, and in the majority of such stories the United States is one of the nations involved. These stories represent a third general category of international news since they reflect reporting of events that are somehow multinational in character.

The total number of foreign nations mentioned in international news stories ranged from 1 to 15. Perhaps a more meaningful statistic is the mean number of foreign nations mentioned in a typical weeknight news broadcast. That average was approximately 12 for each network.

In summary, international news on the U.S. television networks is quite strongly focused on affairs that somehow involve the United States, as indicated by reference to the U.S. in almost 60 percent of all sampled stories. That percentage is divided about equally into stories reporting bilateral affairs of the U.S. and other nations, and news of multilateral affairs. Approximately 25 percent of all international news involves only a single foreign nation, and the remaining 15 percent reports the affairs of two or more foreign nations.

WORLD NEWS LEADERS DURING 1972-1981

A better picture of network television's news geography can be gained by examining the particular nations which were covered by the networks during the 1972–1981 period. As indicated earlier, two types of content data may be used to assess coverage. The first is data concerning those nations which were most frequently mentioned in sampled news stories. The second is data about the nations from which foreign video reports originated. The following pages discuss both types of evidence.

Nations Most Frequently Referred to in the News

Table 3.2 shows the extent of coverage given to more than 50 nations most frequently referred to in the news. The table ranks each nation according to the extent of coverage on each network, with the nations ordered according to their ranking on ABC news. From this table, it is possible to identify several important aspects of network television's news geography for the 10 years from 1972 through 1981.

First, the table shows clearly that the overall pattern of network news attention is highly skewed toward a small number of nations. Only about

Table 3.2
Coverage of 50 Most Frequently Mentioned Nations, 1972-1981,
Expressed as a Percentage of Sampled International Stories

Nation	ABC		CBS		NBC	
	Rank	% of Stories	Rank	% of Stories	Rank	% of Stories
United States	1	57.0	1	60.5	1	58.8
USSR	2	16.7	2	17.1	2	16.2
Israel	3	14.3	3	13.4	3	13.6
Britain [a]	4	9.8	4	9.9	5	8.8
South Vietnam [b]	5	9.1	5	8.7	4	9.0
Iran	6	8.7	6	8.5	8	7.4
Egypt	7	7.7	7	7.7	6	8.0
North Vietnam	8	7.5	9	7.1	7	7.8
France	9	6.2	8	7.2	9	6.3
China, People's Republic	10	5.3	10	4.6	10	4.6
Lebanon	11	4.2	12	3.9	17	3.0
West Germany	12	4.1	11	4.4	11	4.0
Japan	13	4.1	14	3.4	15	3.2
Syria	14	3.5	13	3.5	13	3.2

Table 3.2 continued on next page

Table 3.2 (continued)

Nation	ABC		CBS		NBC	
	Rank	% of Stories	Rank	% of Stories	Rank	% of Stories
Cuba	15	3.2	15	3.2	14	3.2
Poland	16	3.1	17	2.9	17	3.0
Saudi Arabia	17	2.9	13	3.5	12	3.3
Italy	18	2.8	20	2.3	16	3.1
Kampuchea [c]	18	2.8	16	3.1	18	3.0
Afghanistan	19	2.3	29	1.4	23	1.5
Zimbabwe [d]	20	2.2	21	2.2	25	1.3
South Africa	21	2.2	18	2.7	22	1.7
Northern Ireland	22	2.1	26	1.6	23	1.5
Canada	23	1.8	19	2.6	19	2.3
Iraq	24	1.8	30	1.3	27	1.2
Turkey	25	1.6	27	1.5	27	1.2
Jordan	25	1.6	24	1.8	26	1.3
Switzerland	26	1.4	19	2.6	24	1.4
Libya	26	1.4	25	1.7	26	1.3
Mexico	27	1.4	23	1.9	21	1.8
South Korea	28	1.3	22	1.9	20	2.1
India	29	1.3	31	1.1	28	1.1
Spain	30	1.2	26	1.6	25	1.3
Pakistan	30	1.2	35	0.9	33	0.8
Panama	30	1.2	28	1.4	35	0.7
Cyprus	31	1.1	33	1.0	31	0.9
Greece	31	1.1	30	1.3	25	1.3
The Philippines	32	1.0	36	0.8	34	0.8
Thailand	33	1.0	34	1.0	24	1.4
The Vatican	33	1.0	35	0.9	29	1.1
The Netherlands	34	0.9	37	0.8	35	0.7
Algeria	34	0.9	35	0.9	36	0.7
Angola	34	0.9	35	0.9	34	0.8
Laos	34	0.9	41	0.6	30	1.0
Uganda	34	0.9	35	0.9	32	0.9
Portugal	35	0.8	32	1.1	30	1.0
Austria	36	0.8	40	0.7	37	0.7
Argentina	37	0.8	37	0.8	38	0.6
Nicaragua	38	0.7	40	0.7	37	0.7
Chile	38	0.7	34	1.0	34	0.8
Sweden	38	0.7	33	1.0	38	0.6
East Germany	38	0.7	38	0.8	33	0.8
Taiwan	39	0.6	39	0.7	34	0.8
Belgium	40	0.6	40	0.7	33	0.8
N = (stories)		2377		2391		2286

[a] Excludes Northern Ireland.
[b] After the year 1976, all references to Vietnam were coded as North Vietnam.
[c] Formerly Cambodia.
[d] Formerly Rhodesia.

Note: Rankings are based on the absolute number of stories in which each nation was mentioned. Due to rounding, nations with different ranks may appear to be cited in the same percentage of sampled stories. Nations are listed in the order of their rank on ABC. More than 50 nations are included in the table because of differences across networks. Percentages sum to more than 100 percent because multiple nations may be mentioned in a single news story.

25 to 30 nations are mentioned in 2 percent or more of the international news items sampled for any network. Therefore, in relative terms, a nation involved in more than 2 percent of international news on any network received "extensive" coverage.

Second, a large number of the nations that appear most frequently on the network news were involved in wars or major conflicts during the time period studied. Notably, they include all of the combatants in the Vietnam war and the 1973 Middle East war.

Third, the most frequently mentioned nations include such world powers, politically and economically, as the USSR, Great Britain, France, the People's Republic of China, West Germany, and Japan. The one foreign nation appearing most frequently on the network news is the USSR, reflecting its global involvement in international affairs and the often observed tendency of the U.S. networks to interpret events in an East–West or Us-versus-Them framework.

Fourth, Table 3.2 shows the basic similarity of international news coverage by the three major U.S. television networks. The rank order of nations according to the amount of coverage each received is very similar across the networks. The USSR ranks first, Israel second, and South Vietnam third on all three networks. Also, a glance at the ranks of the 15 nations most frequently mentioned on each network shows no major discrepancies in rank ordering across the three networks. This same similarity in coverage patterns was explored in some detail in an earlier cross-network comparison based on 5 of the 10 years covered in this study (Larson, 1978).

Finally, it is worthwhile noting what Table 3.2 does not contain. It provides no indication of how television coverage shifted over time.[2] Some of the news leaders appear in the table because of brief periods of intensive coverage, for example South Vietnam, Afghanistan, and Cyprus. Other nations received more consistent levels of coverage over the entire decade, for example Britain, France and Japan. The table also fails to convey any information about the formats in which news of each nation was presented or the thematic content of such news. Dimensions like these will be dealt with on a region-by-region basis in the following sections of the chapter.

Figure 3.1 is a shaded map of the world which visually represents some of the preceding findings. Nations have been placed in three categories and shaded accordingly. Those nations mentioned in over 5 percent of sampled stories received high coverage, those mentioned in 2 to 5 percent of the sample are in the medium coverage group, and all other nations are in the low coverage group.

---

[2] Appendix F contains data arranged by year for more than 50 most frequently-mentioned nations on each network. It is an expansion of Table 3.2 to include data for each of the 10 years in the sample.

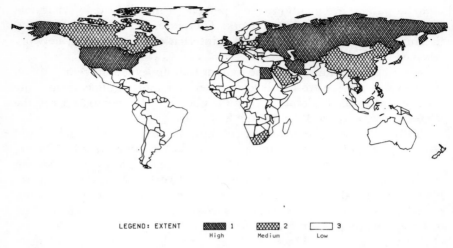

LEGEND: EXTENT   ▨▨▨ 1   ▨▨▨ 2   ☐ 3
                   High   Medium   Low

Figure 3.1
Amount of Coverage by Nation 1972-1981

Since Table 3.2 is based on the number of news story references to nations, it does not provide a good indicator of the independent newsgathering overseas by the networks. The following section assesses those efforts
by looking at the nations from which foreign video reports originated during the 1972–1981 period.

National Origin of Foreign Video Reports Broadcast by ABC and CBS

Table 3.3 shows the 25 nations from which most foreign video reports
originated on the early weeknight news broadcasts of ABC and CBS.[3] The
table is instructive because it represents a direct measure of newsgathering conducted directly by the networks. As noted earlier, each foreign
video report represents a commitment of manpower and money to report
the news, hopefully on a very timely basis and with as much visual content as possible. In some respects Table 3.3 is similar to Table 3.2 and in
other important ways it is different.

First, Table 3.3, like Table 3.2, is skewed toward coverage of a few nations. However, it is much more highly skewed than was Table 3.2. More
than one-quarter of all foreign video reports in the 10-year sample originated directly from two nations, Great Britain and Israel. Add the five
next highest ranking nations and those eight countries represent the point

---

[3] A full 10 years of sample data on the origin of foreign video reports is available only for
ABC and CBS.

Table 3.3
Nations From Which Foreign Video Reports Most Frequently Originated
on ABC and CBS Networks, 1972-1981, Expressed as a Percentage of
All Foreign Video Reports Sampled for Both Networks

Nation	Rank	% of Sampled Foreign Video Reports	Cumulative Percent
Great Britain[a]*	1	17.6	17.6
Israel*	2	7.7	25.3
Vietnam[b]*	3	5.7	31.0
Egypt*	4	4.9	35.9
France*	5	4.6	40.5
Iran	6	4.4	44.9
West Germany*	7	4.1	49.0
Lebanon*	8	3.8	52.8
USSR*	8	3.8	56.6
Poland	9	3.4	60.0
Japan*	10	3.0	63.0
Italy*	11	2.3	65.3
Northern Ireland	12	2.2	67.5
Zimbabwe[c]	13	1.9	69.4
China, People's Republic	14	1.8	71.2
Canada	15	1.6	72.8
South Africa	16	1.4	74.2
Switzerland	16	1.4	75.6
Syria	16	1.4	77.0
Kampuchea[d]	17	1.1	78.1
The Philippines	18	0.9	79.0
Hong Kong*	19	0.8	79.8
Jordan	19	0.8	80.6
Nicaragua	19	0.8	81.4
Portugal	20	0.8	82.2
N = (stories)		1537	

[a] Excludes Northern Ireland.
[b] Includes foreign video reports originating from both North and South Vietnam.
[c] Formerly Rhodesia.
[d] Formerly Cambodia.

Note: The nations marked with an asterisk all had permanent network bureaus during the entire decade, except for Vietnam.

of origin for more than half of all sampled foreign video reports. All 25 nations included in the table account for over 80 percent, an overwhelming proportion of such reports.

Second, although Table 3.3 includes many nations that were involved in war or other major conflicts, some such nations, including Cuba, Afghanistan, and Iraq rank much lower than they did in Table 3.2, based on references in the news. In the case of Cuba, its involvement in war was in Angola, and foreign video reports would likely originate from the latter nation. Afghanistan and Iraq both present difficulties for network correspondents and camera crews, a factor which may largely explain the differences in their rankings.

Third, Table 3.3 includes major economic and political powers or allies of the United States. While this is to be expected, it is noteworthy that the USSR ranks much lower as a point of origination for correspondent's reports than it does as a nation mentioned on U.S. network news. The most plausible explanation for this difference would appear to be Soviet restrictions on the newsgathering activities of network television correspondents in the USSR.

Fourth, there are nations in Table 3.3., such as Nicaragua and Portugal, which rank higher in terms of direct foreign video coverage than they do in terms of news item references. Together with the other patterns noted in Tables 3.2 and 3.3, this would seem to suggest that factors such as cost and lack of political obstacles to newsgathering are important influences on the network news organizations. This importance of economic, technical and logistical factors can also be seen by looking at those nations with permanent network bureaus during the 1972–1981 decade. The 10 nations with bureaus for the entire 10 years accounted for 52.6 percent of all foreign video reports sampled. Adding Vietnam, which had network bureaus during the years it was heavily covered, raises the proportion to 58.3 percent. The influence of network bureau location on television news coverage is a topic that will be explored in more detail in Chapter 5.

Finally, it should be noted that the addition of data for NBC news would not change the pattern in any substantial way. Analysis of NBC data for the five years from 1977 through 1982 show that cross-network differences are minimal. There were no major shifts in the rank order of nations from network to network on this dimension of content.

The pattern of foreign video coverage presented in Table 3.3 and discussed above is represented visually in Figure 3.2. The 25 shaded nations are those which accounted for 82 percent of all foreign video reports broadcast by the networks between 1972 and 1981.

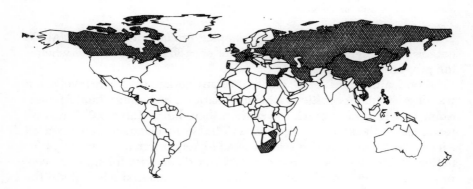

Figure 3.2
The Origin of 82 Percent of Foreign Video Reports on ABC and CBS 1972-1981

COVERAGE OF MAJOR GEOGRAPHICAL REGIONS

What was the pattern of network television coverage of major geographical or geopolitical regions of the world during the 1972–1981 decade? The answer to this question is important for several reasons. In some cases major geographical regions reflect groups of nations with historic cultural, linguistic or political ties. Such ties may relate directly to the growth and development as well as the current status of the international news system in a region and its links to other parts of the world. For example, sub-Saharan Africa is a region that was colonized primarily by France and Britain so it is hardly surprising that Agence France Presse (AFP) and Reuters are today the dominant transnational news agencies in that region. On the other hand, Latin America has historically been more of an American sphere of influence, so that AP and UPI are relatively stronger there. In the case of Eastern Europe, which became part of the Soviet bloc following World War II, there are geopolitical reasons why most of those nations are not extensively covered by major U.S. media.

Because of the economic, political, cultural and other affinities of nations within major regional groups, and the relationship of those ties to the international news communication structure, a region-by-region analysis, as well as a careful look at network news coverage within regions, is important. For a more practical reason, because of the sheer number of nations in the world, regional analysis should help the reader to better understand some of the major patterns and trends in television coverage by the U.S. networks.

In order to examine television's news geography in more detail, nations were aggregated into six geographical regions, not including Canada, approximating those used in official publications of the United Nations and the World Bank. Appendix D contains a list of all nations and territories in each region. Some of the regional groupings are based on political as well as geographical considerations. For example, the Soviet Union, which straddles both Europe and Asia, is grouped with the nations of Eastern Europe which comprise a Soviet sphere of political influence. Also, the nations of the Pacific are grouped with those of Southeast Asia partly because of the association of the Philippines and Guam with news coverage of the Vietnam conflict.

The following sections of this chapter describe coverage of the world's major geographical regions on two levels. The first is an inter-regional comparison which examines the extent and nature of coverage across each of the six regions. The second level focuses on intra-regional patterns and the particular nations and events that comprise most of the news reported. At each level, four important dimensions of coverage are described:

1.   the overall amount of coverage during the decade,
2.   the formats in which it was presented,

3. its thematic content, and
4. changes in coverage over time.

Interregional Comparison of Coverage, 1972-1981

Table 3.4 describes the proportion of all international news involving each of the six geographical regions and Canada during the 1972–1981 period. The relative emphasis given to each region may be quantified in two ways. First, since international news items often refer to nations in more than one region of the world, the percentage of story units referring to all regions of the world sums to more than 100 percent. The second approach is to quantify emphasis by computing the percentage of all national references accounted for by each world region. This has been done for all networks combined in the final column of Table 3.4. With such an approach the percentages do sum to 100 percent. The relative emphasis given to each major region is very similar with either approach to quantification, and there is no difference in the resulting rank ordering of regions.

The relative proportion of coverage given each region appears to place them in three groups, with Western Europe, Asia and the Middle East on top with references in approximately 30 percent of sampled stories, Eastern Europe (including the USSR) in the middle with references in about 20 percent of the stories, and Latin America and Africa on the bottom with about 11 percent and 7 percent respectively. If the nations of Eastern Europe were considered separately from the USSR, they would receive less coverage than any of the other regions, with the exception of Canada, which is the only North American nation other than the United States.

Measured in terms of their proportion of all references in the news, three regions—Western Europe, the Middle East, and Asia—account for more than two-thirds of network news attention during the 1972–1981 period. With the addition of the USSR and Eastern Europe, 84 percent of all coverage is accounted for. Also, as shown in the final column of Table 3.4, Latin America and Africa combined comprise less than 15 percent of all national references on the network newscasts. What this clearly indicates is the network television news "blind spots" during the 10 years from 1972–1982. Except for major crises, the U.S. television networks have paid minimal attention to Latin America, Africa, South Asia, Southeast Asia following the Vietnam war, and Eastern Europe if the USSR is excluded. With the exception of the Eastern European nations, these blind spots comprise a major portion of the Third World. Chapter 4 will build on this finding and examine the coverage of developed, developing and socialist nations in more depth.

Report Formats

As indicated in Chapter 2, each of the three basic story formats used by the U.S. television networks implies something about the reporting and

Table 3.4

Proportion (percentage) of National References by World Region

Region	N = (nations)	ABC	CBS	NBC	All Networks	All References
Western Europe	27	29.6	32.2	30.3	30.7	23.8
Middle East	20	30.5	28.5	28.9	29.3	22.7
Asia	33	29.0	26.6	28.7	28.1	21.8
Eastern Europe [a]	9	20.3	20.5	19.8	20.2	15.7
Latin America	34	11.1	11.5	10.8	11.1	8.6
Africa	43	7.6	8.2	5.7	7.2	5.6
Canada	1	1.8	2.6	2.4	2.3	1.8
N = (stories)		2331	2341	2212	6884 N = (references)	8871

[a] Including the USSR.

Note: The first four columns of percentages add to more than 100 percent due to multiple references to countries across regions in some news stories. The final column (All References) sums to 100 percent.

newsgathering process. In general, anchor reports indicate a reliance on the major international news agencies; domestic video reports most often reflect the Washington D.C. perspective of U.S. government involvement with events; and foreign video reports represent the greatest commitment of network resources in coverage of international affairs.

Table 3.5 shows, for each major world region and Canada, the proportion of network television coverage constituted by each major story format. Several important patterns are displayed in the table.

First, it shows some regional differences in the extent to which network television relies on international news agencies for coverage of international affairs. Overall, for the entire 1972–1981 decade, approximately 41 percent of all international news stories were anchor reports, the vast majority of which came directly from the news agencies. However, nearly half of African news came in that format, suggesting a heavier reliance on news agencies for coverage of that continent.[4] On the other extreme, only a third of Middle East news consisted of anchor reports, indicating a lower than average reliance on the major wire services.

Second, Table 3.5 reveals some differences in the extent to which major regions are covered in the form of domestic video reports. Just over one-fourth of all international news broadcast by the networks during 1972–1981 consisted of domestic video reports. As a region, Latin America exceeded this average, with more than 36 percent of its coverage coming in the form of domestic video reports. This pattern may be partly explained by the geographical proximity of Latin America to the United States and by U.S. foreign policy interests such as the Panama Canal treaty negotiations and regular noncrisis interaction with Mexico and Cuba. Asia also had a higher than average proportion of coverage in the form of domestic video reports, reflecting direct U.S. concern with events such as the Vietnam peace negotiations, Indochinese refugee resettlement, trade with Japan, and relations with China. Eastern Europe's higher than average percentage of domestic video report references largely reflects Washington coverage involving the Soviet Union. On the other hand, a low proportion of Western European and African coverage consisted of domestic video reports. While this may reflect Africa's relatively low priority on the U.S. foreign policy agenda, such an explanation is not applicable to Western Europe. Instead, European coverage appears to result from the concentrated presence of both news agency and network television bureaus in the region. Such a concentration increases the likelihood that news from the region will be reported either directly by a network correspondent, or in the form of an anchor report drawing on news agency output.

Third, Table 3.5 indicates some noteworthy differences by region in foreign video report coverage by the networks. During the decade of this

---

[4] For the purpose of this study, Africa refers to sub-Saharan Africa.

Table 3.5

Percentage of References in Each Major Story Format by World Region

Region	Anchor Report	Domestic Video Report	Video Report	$N =$ (stories) [a]
Western Europe	40.3	16.1	43.6	2113
Middle East	33.3	29.5	37.2	2018
Asia	41.7	31.1	27.3	1932
Eastern Europe	38.7	31.6	29.7	1391
Latin America	37.7	36.4	25.8	766
Africa	47.4	20.0	32.6	494
Canada	31.8	31.8	36.3	157
World Average	41.4	25.9	32.7	6884

[a] The final column represents the base N for each row in the table. It is the total number of stories in which each region was mentioned. Column percentages and N are not meaningful since more than one region may be mentioned per news story.

study, just under one-third of all international news broadcast by the three networks consisted of foreign video reports. As shown in the table, a higher proportion of such reports referred to nations of Western Europe and the Middle East and a lower proportion referred to countries of Latin America. However, the foreign video report column of Table 3.5 should be interpreted with caution since it only indicates *the percentage of such reports that mention each region, not the regions from which such reports originate.* Data on the origination of foreign video reports were collected for two of the three networks, ABC and CBS and are presented in Table 3.6.[5] Table 3.6 shows the proportion of all foreign video reports broadcast by ABC and CBS during 1972–1981 which originated in each of the major geographical regions and Canada. Based on NBC data from the 1976–1981 period, it appears that the pattern reflected in this table would change very little if the third network were included. The regions are included in the table in rank order of their importance, based on the number of foreign video reports originating in each region.

Table 3.6
Origin of Foreign Video Reports (percentage) on ABC and CBS by Region, 1972-1981

Region	ABC	CBS	Both Networks[a]
Western Europe	41.6	35.8	39.0
Middle East	23.1	25.9	24.3
Asia	15.6	17.4	16.4
Eastern Europe	8.6	7.0	7.9
Latin America	5.4	5.9	5.6
Africa	5.3	5.2	5.2
Canada	0.6	2.7	1.8
N = (stories)	841	698	1539

[a] Column percentages may not sum to 100 percent because of rounding error.

As Table 3.6 indicates, Western Europe was the region covered most intensively by the networks through the use of their own correspondents, with nearly 40 percent of all foreign video reports originating from that region. Without question, the presence of permanent bureaus in a number of European cities facilitated this coverage. During the 1972–1981 period both ABC and CBS maintained bureaus in Bonn, London, Paris and Rome.

Following Western Europe, the networks focused the newsgathering efforts of their own correspondents most heavily on the Middle East, which accounted for about one-quarter of the foreign video reports sampled. As with Europe, the predominance of the Middle East relates at least in part

---

[5] For NBC, data on the origination of foreign video reports were collected for the years 1976–1981.

to the presence of permanent network bureaus in the region. During 1972–1981 ABC and CBS had bureaus in Beirut, Cairo and Tel Aviv.

The third-ranking region in terms of foreign video report coverage was Asia, which made up over 16 percent of such coverage by ABC and CBS combined. Like Western Europe and the Middle East, there were network bureaus in Asia during 1972–1981. ABC and CBS had bureaus in Tokyo, Hong Kong and Vietnam during the war years.

The remaining foreign video report coverage by ABC and CBS was divided among Eastern Europe (including the USSR), Africa, Latin America, and Canada. The low levels of coverage given to these remaining regions shows, more dramatically than Table 3.5, the newsgathering priorities set by the U.S. television networks.

In summary, given a limited number of correspondents, bureaus, and money with which to work, the networks deploy their resources as follows. Nearly two-thirds of all coverage originating from network correspondents overseas comes from two regions, Western Europe and the Middle East. Adding Asia accounts for 80 percent of all foreign video reports broadcast by CBS and ABC during 1972–1981, principally because of intensive coverage of the war in Indochina. The remaining 20 percent of direct visual coverage from overseas is distributed among Eastern Europe, Latin America, Africa, and Canada. The pattern clearly is one of extreme concentration of network newsgathering resources, principally in Western Europe and the Middle East. The window of network television's news cameras shows a consistent, decade-long pattern of focusing on certain parts of the world at the expense of others. Those others, are predictably the developing nations of Latin America, Africa and, since the end of the Vietnam War, South and Southeast Asia as well. The importance of this finding will be stressed again as part of the analysis in the following chapter.

Trends Over Time

Trends in coverage of the major geographical regions differ considerably over the 10 years from 1972 through 1981. Table 3.7 shows the proportion of international news stories mentioning each geographical region and Canada by year. Based on the information in this table, the regions may be thought of in three categories.

First, there are two regions that received relatively consistent or continuous levels of coverage over the 10-year period. They are Western Europe and Eastern Europe, including the Soviet Union. Western Europe's proportion of international news on the networks ranged from 20 to 41 percent and was in the 30 to 40 percent range during all but 3 of the 10 years. The consistency or stability of attention to Eastern Europe is largely accounted for by the inclusion of the Soviet Union in that region.

Second, there are a number of world regions that exhibit relatively sharp peaks and valleys in coverage over the decade. These include the

Table 3.7

Percentage of References to Each Major World Region and Canada by Year, All Networks

Region	1972	1973	1974	1975	1976	1977	1978	1979	1980	1981	1972–1981	N
Western Europe	31.1	25.9	41.4	33.1	36.0	39.4	32.5	21.9	19.7	30.9	30.7	2113
The Middle East	8.9	25.1	27.4	23.0	22.6	23.1	26.1	47.2	50.1	34.1	29.3	2018
Asia	65.6	51.5	16.5	34.4	20.0	19.3	16.1	22.3	22.5	12.0	28.1	1932
Eastern Europe	16.5	12.2	22.4	17.3	17.3	16.7	26.9	15.6	26.2	28.8	20.2	1391
Latin America	4.4	6.2	9.4	8.3	12.7	14.2	14.3	15.6	12.7	12.3	11.1	766
Africa	1.5	0.3	2.3	5.2	17.8	17.6	11.9	8.3	3.2	4.4	7.2	494
Canada	1.1	1.7	2.3	1.4	3.6	3.4	3.2	1.7	2.8	2.0	2.3	157
N = (stories)	752	582	532	712	614	654	721	775	785	757		6884

Middle East, Africa and Asia. The proportion of international news coverage involving the Middle East began at a low of 9 percent in 1972. From 1973, the year of the Middle East war and the Arab oil embargo, through 1981, the Middle East was consistently involved in more than 20 percent of international news, peaking in 1979 and 1980 at 47 and 50 percent respectively of international news for those years. Africa received very little coverage at either the beginning or the end of the decade, but attention increased sharply during the 1976–1978 period in response to crises in Angola and Rhodesia (now Zimbabwe). Coverage of Asia by the American television networks was skewed heavily toward the first four years of the 1972–1981 period because of U.S. involvement in Vietnam. The region was involved in nearly two-thirds of all international news in 1972 and over half of world news in 1973. Following U.S. withdrawal from Vietnam in 1975, overall coverage of the region dropped to a relatively low level.

Third, there is one region of the world, Latin America, which shows a gradual increase in network television news coverage over the decade. The increase in that region may be largely explained by trends in coverage of just two nations, Cuba and Mexico (McAnany, Larson, and Storey, 1982). Excluding the Soviet Union, Eastern Europe also would show increased levels of coverage during the second half of the decade.

Thematic Content

The thematic content of television news also varied from region to region during the 1972–1981 decade. Using the *Television News Index and Abstracts*, news stories were coded into nine thematic categories. Four of the news themes can be considered crisis themes and five noncrisis themes as follows:

Crisis Themes
1. Unrest and Dissent
2. War, Terrorism, and Violent Crime
3. Coups and Assassinations
4. Disasters
Noncrisis Themes
5. Political-Military Affairs
6. Economics
7. Environment
8. Science-Technology
9. Human interest

Table 3.8 shows the proportion of stories dealing with crisis themes for each region of the world and Canada. According to the table, network

Table 3.8
Proportion of Stories With Crisis Themes, by World Region, 1976-1981 [a]

	ABC		CBS		NBC		All Networks	
Region	N	%	N	%	N	%	N	%
Western Europe	443	28.9	444	29.5	385	24.2	1272	27.7
Middle East	529	26.5	507	25.8	459	24.6	1495	25.7
Asia	294	25.9	260	30.8	252	26.6	806	27.7
Eastern Europe	334	19.5	326	18.4	294	16.7	954	18.2
Latin America	204	30.4	204	31.4	180	25.6	588	29.3
Africa	157	38.2	165	35.8	110	40.0	432	37.7
Canada	30	30.0	48	20.8	39	25.6	117	24.8
World Total	1484	28.1	1476	28.1	1346	25.0	4306	27.1

[a] Data on story themes were gathered for all three networks only for the 1976–1981 period.

Note: The percentage columns represent the proportion of crisis themes, computed on the base number (N) of stories to the left of each percentage figure.

television's coverage of crisis news is quite evenly distributed across the major regions of the world, with two notable exceptions. The first is Africa, which had a measurably higher proportion of crisis news than other world regions. The high proportion reflects intensive coverage of the major crises in Angola and Rhodesia, but it also points to the dearth of news from Africa in the absence of crises. The data indicate that, more than any other world region, Africa tends to be ignored by the U.S. television networks except in the case of a war or crisis of major proportions.

A second exception to the general pattern shown in Table 3.8 is the relatively low proportion of crisis news from Eastern Europe. This might be explained in part by the degree of control exercised by governments over Western reporters, as well as different standards for judging news values which de-emphasize violence and unrest.

The Middle East might have been expected to have a higher ratio of crisis to noncrisis news than indicated in Table 3.8. However, a great deal of news from that region dealt with diplomatic efforts to find solutions to the Arab–Israeli conflict, rather than coverage of the conflict *per se.*

A similar situation pertains in the Asian region where a higher proportion of crisis news might have been expected because of the Vietnam war. However, a great deal of the coverage in this period dealt with negotiations or diplomacy rather than coverage of the war itself. Also, coverage of crisis news is distributed fairly evenly over the three major Asian subregions, Southeast Asia, East Asia, and South Asia. It included coverage of political unrest in Korea, China and India, war in Afghanistan, and earthquakes in China and Japan (Larson and Storey, 1983).

The preceding description of network television's pattern of news attention during the 1972–1981 period compared major geographical regions of the world in terms of the extent of their television coverage, its formats, changes in coverage over time, and the thematic content of such coverage. Following sections of this chapter will discuss the same dimensions of coverage but will focus on the particular nations and events covered within each of the major regions during the decade.

Coverage Within the Major Regions, 1972-1981

In order to describe international news coverage within each of the major geographical regions, attention will be focused primarily on the news leaders, those nations which accounted for the bulk of network news coverage during the 1972–1981 decade. The following description treats the regions in order of their emphasis by the networks, beginning with Western Europe, the region receiving the most coverage during the decade, and ending with Africa, the least-covered region of the world.

Western Europe

Nearly all of the 27 nations in Western Europe, including Scandinavia, were referred to at least once in the news broadcasts sampled for this study. The only exceptions were small nations, such as Andorra and Liechtenstein. However, there were important differences in the extent and nature of coverage given to the nations of the region. Table 3.9 rank orders the top 15 news leaders of Western Europe and provides information concerning the extent of coverage, story formats, and thematic content of the coverage.

Not surprisingly, the top four ranking nations in Western Europe were also those with permanent network bureaus during the entire 1972–1981 period. Taken together, Britain, France, West Germany, and Italy accounted for approximately 60 percent of all the region's news coverage, measured as a proportion of total references to Western European nations in the news. All 15 nations in the table account for more than 90 percent of all news from Western Europe. In other words, about half of the nations in the region account for nearly all of its coverage on network television news. The remaining coverage, less than 10 percent of the region's total, was divided among the other nations, principally Belgium, Denmark, Iceland, Ireland, Norway, and Finland.

Report Formats As already indicated, the networks place a high priority on Western Europe in the allocation of their newsgathering resources. Seven nations in Table 3.9 were also listed earlier in Table 3.3 along with 18 other nations from which foreign video reports were most likely to originate. In addition, Table 3.9 shows that a number of Western European nations received a large proportion of all their references in foreign video reports. For each of the top six nations in the table, 45 percent or more of

Table 3.9
Coverage Given the News Leaders of Western Europe, 1972-1981

Nation	N = (references)	% of References[a]	Story Format[b]			Crisis Themes[c]	
			Anchor %	Domestic %	Foreign %	%	N
1. Great Britain	670	24.7	39.3	10.9	49.9	25.4	236
2. France	463	17.1	36.3	19.2	44.5	25.0	172
3. West Germany	293	10.8	29.7	23.9	46.4	32.7	104
4. Italy	194	7.1	35.1	14.9	50.0	48.2	56
5. Switzerland	128	4.7	29.7	23.4	46.9	13.1	61
6. Northern Ireland	121	4.5	38.0	1.7	60.3	81.1	37
7. Turkey	99	3.6	40.4	27.3	32.3	51.5	33
8. Spain	95	3.5	47.4	21.1	31.6	51.4	37
9. Greece	87	3.2	46.0	16.1	37.9	54.8	31
10. Cyprus	71	2.6	40.8	19.7	39.4	37.5	24
11. The Vatican	68	2.5	30.9	8.8	60.2	28.6	21
12. Portugal	68	2.5	52.9	8.8	38.2	46.2	26
13. The Netherlands	58	2.1	48.3	8.6	43.1	78.9	19
14. Sweden	54	2.0	55.6	20.4	24.1	12.5	24
15. Austria	50	1.8	30.0	14.0	56.0	25.0	16
N = (references)	2711						

[a] The base for these percentages is 2711, the total number of references to all nations of Western Europe, including some not in the above table.
[b] The base number for each row of story format percentages is the number of story references for each particular nation, found in the first column of the table. Percentages for the three story formats may not add to 100 percent due to rounding error.
[c] The proportion of crisis themes for each nation is based on data for only the "CBS Evening News," 1972–1981.

their mentions came during foreign video reports. Two nations, Northern Ireland and the Vatican, received more than 60 percent of their mentions in the foreign video format. However, it should be stressed that these data indicate only references to nations in foreign video reports, not the national origin of such reports. Only in the case of Great Britain does the proportion of video reports originating from that nation approximate the 49.9 percent figure in Table 3.9. For example, many of the foreign video reports that refer to the Vatican originated from Italy and in some cases other European countries. Likewise, some of the reports mentioning Northern Ireland originated in London, and so forth. In summary, the data in Table 3.9 provide evidence of heavy use of foreign video reporting by the networks in covering Western Europe, but data concerning the origin of such reports reveals a pattern in which certain nations, particularly those with permanent network bureaus, predominate.

In general, the news leaders of Western Europe receive a low proportion of their references in domestic video reports. Hardly any of the news from Northern Ireland was delivered in this format, and the Vatican, Portugal and the Netherlands are all extremely low on this measure. News from these four nations and territories tended not to involve the U.S. government officially as a participant in diplomacy or in some policy role. To use the example of Northern Ireland, the rioting and bloodshed there made for visually exciting and newsworthy television according to present news selection criteria, but the U.S. government was never portrayed as part of the problem or the solution to that civil conflict.

On the other hand, news from Turkey and West Germany, both important NATO allies of the United States, showed the highest proportion of references in domestic video reports. Switzerland, Sweden and Spain also showed a relatively high proportion of references in the domestic video format.

For the most part, Western European news leaders appeared a high proportion of the time in anchor reports, for which the networks depend heavily on the international news agencies. Sweden, along with the other Scandinavian nations, is especially high in this regard. The proportion of references in anchor reports is 64 percent for Finland, 46 percent for Norway, and 48 percent for Denmark. This greater degree of reliance on the news agencies for coverage of Scandinavia may relate in part to bureau locations. However, the size and political importance of these nations may also be part of the explanation.

Trends Over Time The preceding discussion presented a static picture of how leading West European nations in the news were covered during 1972–1981. However, examination of the data on a year-by-year basis shows some differences in the pattern of network attention to West European nations over the 10-year time period. As might be expected, the top five news leaders from the region, along with a number of other nations received some coverage during each of the 10 years. Although the intensity of coverage varies and there are some fluctuations in coverage of each nation from year to year, the overall pattern is one of relatively consistent coverage of a large number of West European nations.

The overall pattern of consistency does not hold for all nations in the region. There were three major exceptions. First, two nations which showed concentrated periods of intensive coverage only during crises were Cyprus and Portugal. The networks focused their greatest attention on Cyprus in 1974, when it was involved in a civil war. Coverage of Greece and Turkey, the other two nations with direct involvement in that conflict, also peaked during 1974. Portugal received its heaviest coverage during 1975, a year which encompassed anti-Nato demonstrations, an attempted violent military coup, parliamentary elections, and the granting of independence to Angola, which was followed almost immediately by a civil war in that nation.

A second example of dramatic change in levels of coverage during the decade was the Vatican. It had received only sporadic network coverage during the six years from 1972 through 1977. That pattern changed abruptly in August of 1978 with the death of Pope Paul VI. Preparations for and coverage of his funeral were followed by attention to the selection of a new Pope, John Paul I, by the college of cardinals. A similar cycle of coverage began again just over one month later with the sudden death of the new Pope. In October, the networks gave heavy coverage to the selection of a new Pope, Karol Cardinal Wojtyla of Poland, who took the name of Pope John Paul II. Since then, the worldwide travels of this new Pope and the assassination attempt in which he was wounded have generated considerable coverage by the networks. The activities of the Pope have become a familiar part of the network news menu.

Finally, there is a group of nations which received low levels of coverage on and off during the decade. It includes Belgium, Finland, Denmark, Greece, Iceland, Ireland, and some of the very smallest West European nations.

Thematic Content The final column of Table 3.9 shows the proportion of crisis themes in "CBS Evening News" stories about each of the news leaders during the 1972–1981 period.[6] It reveals quite graphically the "hot spots" in Western Europe during the decade of this study. Approximately 80 percent of the stories that referred to Northern Ireland and the Nether-

---

[6] Data concerning story themes was gathered for the entire 1972–1981 period only for CBS News. Thematic data for the other two networks is available for the 6 years from 1976–1982. However, it is not presented in Table 3.9 in order to avoid weighting the proportions more heavily for events covered during the final 6 years of the decade. This procedure will be followed in the remaining tables of this chapter.

lands dealt with crisis themes. Network television viewers during this period are familiar with the violence between Protestants and Catholics in Northern Ireland. In the Netherlands, some of the crisis coverage was due to terrorist incidents. For example, there was intensive coverage of two incidents in December 1975 in which terrorists from South Molucca Island in Indonesia, a former Dutch colony, hijacked a train and took over the Indonesian consulate in the Netherlands. Hostages were held for nearly 2 weeks. Five other nations received substantial amounts of crisis coverage, ranging from 45 to 55 percent of all stories referring to them on television news. They were Italy, Turkey, Spain, Greece, and Portugal. Each of these nations was involved in some form of war, terrorism, or civil strife during the 1970s, and each of those crises drew the attention of the U.S. television networks, sometimes nearly the only significant attention given to a nation.

The Middle East

All 20 nations in the Middle East, which includes the northern tier of Africa, appeared in the network news during the 1972–1981 period. However, just half of those nations, the 10 included in Table 3.10, accounted for more than 95 percent of all references to the region.

The closest U.S. ally in the region, Israel, received the most coverage, accounting for nearly 30 percent of all references to nations from the Middle East. That was nearly double the figure for Iran and Egypt, which each accounted for approximately 17 percent of references to Middle Eastern nations. Taken together, these three nations comprise nearly two-thirds of the networks coverage of the Middle East during 1972–1981. As was the case with Western Europe, the presence of network bureaus appears to correlate with the extent of coverage. Both Egypt and Israel had permanent network news bureaus.

A second group of three nations, Lebanon, Syria and Saudi Arabia, each accounted for 7 to 8 percent of references to nations from the region. One nation in this group, Lebanon, also had network bureaus in its capital city, indicating that the presence of a network bureau does not automatically mean that a nation will receive more news coverage than others in its region.

The remaining news leaders from the Middle East include Libya, Jordan, Iraq and Algeria. Each of these nations comprised from 2 to 3 percent of national references in the region.

In summary, the pattern of concentration in network news from the Middle East approximates that of Western Europe. Several nations account for well over half of the coverage and about half of the nations from the region make up more than 90 percent of its news.

Report Formats With three network bureaus in the region, it is hardly surprising that a relatively high proportion of its news should come in the

Table 3.10
Coverage Given the News Leaders of the Middle East, 1972-1981

Nation	N = (references)	% of References [a]	Story Format [b]			Crisis Themes [c]	
			Anchor %	Domestic %	Foreign %	%	N
1. Israel	970	29.3	32.2	28.4	39.5	26.6	320
2. Iran	577	17.4	28.4	35.5	36.0	23.8	202
3. Egypt	550	16.6	26.9	26.2	46.9	17.5	183
4. Lebanon	261	7.9	41.8	7.7	50.6	69.9	93
5. Syria	240	7.2	33.8	12.1	54.2	41.0	83
6. Saudi Arabia	226	6.8	27.9	42.0	30.1	15.7	83
7. Jordan	111	3.3	23.4	27.9	48.6	13.6	44
8. Libya	104	3.1	26.0	48.1	26.0	26.8	41
9. Iraq	96	2.9	30.2	28.1	41.7	32.3	31
10. Algeria	59	1.8	23.7	37.3	39.0	28.6	21
N = (references)	3316						

[a] The base for these percentages is 3316, the total number of references to all nations of the Middle East, including those not presented in this table.

[b] The base number for story format percentages is the number of references to each particular nation, found in the first column of the table. Percentages for the three story formats may not sum to 100 percent due to rounding error.

[c] The proportion of crisis themes for each nation is based on data from the "CBS Evening News" only, 1972–1981.

form of foreign video reports. However, as shown in Table 3.10, there are differences in the extent to which the region's leading news nations appeared in this reporting format. Four nations, Syria, Lebanon, Jordan and Egypt received 47 percent or more of their references in foreign video reports. At the other extreme were Saudi Arabia and Libya which were referred to 30 and 26 percent of the time, respectively, in foreign video reports. Comparing these findings with data on the origin of foreign video reports indicates that most network video reports from the region originated from Israel, followed by Iran, Lebanon, Syria and Jordan in that order. On the other end of the spectrum, two nations, Libya and Saudi Arabia were very seldom the source of foreign video reporting by the networks.

It is probably more than coincidental that Saudi Arabia and Libya, which ranked lowest among the news leaders in foreign video report coverage, rank highest in their proportion of references in domestic video reports. Forty-eight percent of Libya's mentions and 42 percent of the references to Saudi Arabia came in that report format. Presumably this reflects, in both cases, a direct connection between the events being covered and policy concerns of interest in the United States. Concern with international oil prices, the energy problem more generally, and even Billy Carter's relationship with the Libyan government, may all have contributed to this emphasis in the news.

In line with the regional pattern, the leading news nations of the Middle East do not exhibit a high degree of reliance on the international news agencies in their coverage. Approximately one-quarter to one-third of the references to most of the nations are in the form of anchor reports. The one exception is Lebanon, for which the networks show a higher degree of reliance on the news agencies. This pattern may be explained in part by the greater availability and use of video reports, both foreign and domestic, in reporting from these nations.

Trends Over Time The top six news leaders from the Middle East all received some attention from the networks during each of the 10 years from 1972 through 1981. With the exception of a lower level of coverage in 1972, Israel and Egypt generally were covered on a steady and consistent basis throughout the decade. Iran, on the other hand, received very low levels of coverage until the seizure of American hostages at the end of 1979. Its high ranking can be attributed entirely to coverage of the hostage crisis in 1979, 1980, and 1981.

Lebanon, Syria and Saudi Arabia all had periods of relatively intensive coverage during the decade. For Lebanon, the peaks in coverage came during the four years from 1975 through 1978 with increased coverage of its civil war and related internal political problems. Syria received relatively heavy coverage during 1973, the year of war in the Middle East and for the years following that war. Saudi Arabia showed an increase in coverage following 1973, with increased attention to activities of OPEC nations and the world oil crisis. In 1979 coverage of Saudi Arabia again increased, with intensive coverage of an armed takeover of the mosque in Mecca, during November and December of that year. 1981 was also a year of peak coverage of Saudi Arabia, some of it coming in connection with the conflict in Lebanon, and the rest concentrating on Saudi Arabia's foreign relations with a large number of other nations, inside and outside the Middle East.

The remaining news leaders from the region were covered at relatively low levels during most of the decade. Three of the nations, Libya, Iraq and Algeria, showed an increase in coverage near the end of the decade, from 1979 through 1981. In all three cases there appears to be some relationship between the increased coverage and events in Iran during the same period.

Thematic Content Lebanon is the only nation which shows an extremely high proportion of crisis coverage, based on data from the last column of Table 3.10. More than two-thirds of its coverage involved some form of crisis or another. Syria ranks next highest on this dimension, with 41 percent of its coverage involving crisis. The relatively low proportion of crisis themes reported for other nations that were involved in war or crisis during the decade is probably due to the short-lived nature of the conflicts and the heavy coverage of long, drawn out diplomacy between conflicts. A good example would be the "shuttle diplomacy" initiated by Henry Kissinger as U.S. Secretary of State.

Asia

Most of the nations in Asia were mentioned, however briefly, on at least one network news broadcast during 1972–1981. In all, more than 28 of the 34 independent nation states were cited at least once in the sampled newscasts. However, the 10 nations whose coverage is summarized in Table 3.11 accounted for over 90 percent of network television's coverage of Asia.

Vietnam received more than three times as much coverage as the People's Republic of China, which was the second most extensively covered nation in the region. U.S. involvement in the Vietnam war explains that country's position as the news leader in Asia during this time period. Kampuchea (formerly Cambodia), another nation involved in the Indochina conflict, ranks fourth among Asian news leaders. Along with Vietnam and Laos it accounts for over half of all news from the region during 1972–1981, measured as a proportion of total references to nations in the region.

Apart from Vietnam and its neighbors, levels of network television coverage appear to place the news leaders of Asia into two groups. At the

Table 3.11
Coverage Given the News Leaders of Asia, 1972-1981

Nation	N = (references)	% of References[a]	Story Format[b] Anchor %	Domestic %	Foreign %	Crisis Themes[c] %	N
1. Vietnam (North & South)	1154	41.8	35.5	34.5	30.0	38.4	378
2. China, People's Republic	344	12.5	40.7	30.2	29.1	20.7	111
3. Japan	251	9.1	32.7	29.9	37.5	18.3	82
4. Kampuchea	208	7.5	49.5	21.2	29.3	54.8	73
5. South Korea	124	4.5	41.9	38.7	19.4	47.8	46
6. Afghanistan	122	4.4	28.7	33.6	37.7	33.3	33
7. India	82	3.0	50.0	15.9	34.1	29.6	27
8. Thailand	79	2.9	43.0	13.9	43.0	47.8	23
9. Pakistan	68	2.5	44.1	33.8	22.1	42.9	21
10. Philippines	62	2.2	43.5	12.9	43.5	60.0	20
N = (references)	2759						

[a] The base for these percentages is 2759, the total number of references to all nations of Asia, including those not in the above table.

[b] The base number for all story format percentages is the number of story references for each nation, found in the first column of the table. Percentages for the three story formats may not sum to 100 percent due to rounding error.

[c] The proportion of crisis themes for each nation is based on data from the "CBS Evening News," 1972–1981.

top are the People's Republic of China and Japan, which accounted for approximately 13 and 9 percent of all national references respectively. In a second group are nations that account for between 2 and 5 percent of the references to nations of the region. This group includes South Korea, Afghanistan, India, Thailand, Pakistan and the Philippines.

Report Formats Although the proportion of references to all Asian nations in foreign video reports is lower than the overall proportion of such reports broadcast on network television, Table 3.11 shows that several nations are above the overall network average. The presence of network bureaus in Tokyo may account in part for the large percentage of foreign video reports referring to Japan. This may also be part of the explanation for Thailand, which had some network presence for part of the 1972–1981 period. Visually appealing and dramatic events may also be a part of the explanation. Both India and the Philippines experienced repeated natural disasters such as floods, typhoons and earthquakes. Both countries had riots and political demonstrations. The Philippines also had an ongoing war with insurgents in Mindanao. Also, of course, Afghanistan was involved in war late in the decade under study. However, it is well to remember that Table 3.11 reports only references in foreign video reports and does not indicate the proportion of reports originating in each nation.

Asia's diplomatic and strategic importance to the United States is indicated by the use of domestic video reports on network television. Vietnam, China, South Korean, Afghanistan, and Pakistan were highest in domestic video report references. Negotiations over U.S. withdrawal from Vietnam and subsequent talks concerning POWs and MIAs reflect that connection. The normalization of relations with China and speculation about related changes in U.S. foreign policy is another example. Likewise, the U.S. government reacted strongly to events in Afghanistan and began talks with Pakistan about bolstering the defense capability of the subcontinent. All of these issues generated many statements from politicians, government officials, and the White House itself.

On the other extreme, Thailand, the Philippines and India received a low proportion of their references in domestic video reports. Although Bangkok was an important base for correspondents in the early 1970's, neither Thailand nor India were high on the policy agenda in Washington, D.C., during 1972–1981. The low percentage of domestic video reports for the Philippines may reflect the distance the U.S. administration tried to maintain from the Marcos regime during President Carter's human rights campaign. An extension of the content data file through the first two years of the Reagan administration would no doubt show an increased proportion of domestic video reports mentioning the Philippines, if only because of President Marcos' official visit to the United States in 1982.

For coverage of India and Kampuchea, and to a lesser extent Thailand and the Philippines, the networks relied quite heavily on the wire services. The consistently biggest issue out of Thailand in the late 1970's concerned the plight of refugees, which neither generates "interesting" video nor has much bearing on the "important" issues of international politics. News from Kampuchea in the early 1970s consisted mostly of statistics about the bombing runs over North Vietnam and updates on the siege of Phnom Penh. After 1975, the focus was on refugees. None of these issues generated much reaction from U.S. government officials. The generally poor state of U.S. relations with India and the Philippines made coverage of those countries easiest with anchor reports.

Trends Over Time An examination of Asian coverage on a year-by-year basis shows some marked changes in the pattern of network attention to the news leaders of the region. The pattern can be explained by looking at news leaders in each of the major subregions of Asia (Larson and Storey, 1983). Southeast Asia received extremely high levels of coverage in 1972, 1973 and 1975, with almost all of it coming from Vietnam, Kampuchea, Laos, and, to a lesser extent, Thailand.

South Asia received extremely little coverage, with the exception of 1980, following the Soviet invasion of Afghanistan. India did receive some mention throughout the decade, but at a very low level. Pakistan was covered at the end of the period with the attack on the U.S. embassy in that nation.

Although there are some minor fluctuations in coverage of East Asia, it shows the most consistent levels of coverage over the decade. Approximately 10 percent of all network news items broadcast during the period involved one or more East Asian nations, principally the People's Republic of China and Japan, and to a lesser extent, the Republic of Korea.

Thematic Content Table 3.11 also shows how the news leaders of Asia compare on the proportion of crisis coverage each received. South Korea was high on this measure because of an emphasis on civil unrest and human rights violations. Thailand, the Philippines and Pakistan were high because of their proximity to military conflict which spilled across borders or were reported by correspondents safely distant from the combat. Kampuchea had the highest percentage for any nation in which a war was actually fought.

The proportion of crisis themes for the five nations directly involved in warfare (Vietnam, China, Kampuchea, South Korea, and Afghanistan) may appear unusually low. However, diplomatic efforts to resolve the conflicts were often the focus of coverage.

Finally, the low proportion of crisis reporting from Japan is noteworthy. The U.S. television networks depicted Japan as an important trading partner and economic partner of the U.S. In addition, there was considerable network interest in cultural or human interest stories from Japan, perhaps spurred by the presence of permanent bureaus as well as Japan's cultural contrasts with the United States.

Eastern Europe and the USSR

Eastern Europe and the Soviet Union comprise nine socialist nations, all of which were referred to on one or more of the sampled news broadcasts. For all practical purposes, however, only two nations received more than minimal coverage. As shown in Table 3.12, they were the USSR and Poland, which together represent 90 percent of references to nations of the region.

As indicated earlier in this chapter, the USSR is the most frequently mentioned nation on U.S. network television news because of its superpower role and the tendency of U.S. television to report international affairs within an overall framework of East–West conflict. An additional factor increasing the Soviet Union's proportion of East European news is its close political and economic ties with all other nations of the region. Characteristically, the Soviet Union is mentioned in a high proportion of stories involving those other nations.

The Soviet Union aside, Poland is the only East European nation to receive much coverage during the 1972–1981 time period. Its level of coverage was four times that of East Germany, the next most frequently mentioned nation of the region.

Report Formats Although Eastern Europe and the Soviet Union as a region show a low proportion of national references in the foreign video report format, this largely reflects the predominance of data from the Soviet Union. Only about 27 percent of references to that nation occurred in foreign video reports, well under the average for all network international news. However, in the case of Poland, the networks did use their own correspondents. As shown in Table 3.12, nearly 56 percent of references to that nation occured in foreign video reports. Earlier, Table 3.3 showed that Poland ranks ninth in the world as a source of foreign video reporting by the networks. It appears likely that the concentration of network newsgathering efforts on Poland continued into 1982 and 1983, with coverage of martial law, the struggles of the Solidarity labor union, and of course the visit of Pope John Paul II to his home country.

With the exception of the USSR, East European nations tend not to be involved in the networks' domestic video reporting. The implication of this finding is that events in Eastern Europe generally were not a salient part of the policy agenda in Washington, D.C., during 1972–1981.

For the remaining news from Eastern Europe, the networks show their usual dependence on the international news agencies. The proportion of anchor reports in news from Poland is low, but probably only because of the high proportion of foreign video reports from that nation.

Trends Over Time As already indicated, there is an underlying consistency to coverage of Eastern Europe, largely due to the continued high level of attention given the USSR. Considered apart from the USSR the re-

Table 3.12
Coverage Given the News Leaders of Eastern Europe and the USSR, 1972-1981

Nation	N = (references)	% of References[a]	Story Format[b]			Crisis Themes[c]	
			Anchor %	Domestic %	Foreign %	%	N
1. The USSR	1177	76.1	38.7	34.6	26.7	13.0	408
2. Poland	211	13.6	28.0	16.6	55.5	34.8	69
3. East Germany	53	3.4	37.7	17.0	45.3	33.3	18
4. Yugoslavia	47	3.0	42.6	21.3	36.2	33.3	21
5. Czechoslovakia	24	1.6	37.5	16.7	45.8	44.4	9
N = (references)	1546						

[a] The base for these percentages is 1546, the total number of references to the USSR and nations of Eastern Europe, including those not in the above table.

[b] The base number for all story format percentages is the number of story references for each nation, found in the first column of the table. Percentages for the three story formats may not sum to 100 percent due to rounding error.

[c] The proportion of crisis themes for each nation is based on data from the "CBS Evening News" only, 1972–1981.

maining nations of Eastern Europe received increased coverage during the second half of the 1972–1981 decade. Czechoslovakia, East Germany, Poland, and Yugoslavia all received more attention during the last half of the decade than during the first half. The increase was most pronounced in the case of Poland, which received more than half of its coverage in the year 1981 alone. That year included coverage of demonstrations, labor unrest, the Solidarity union, martial law, the church, and also coverage of Pope Paul's activities or statements in relation to his home country. Yugoslavia received most of its coverage during the four years from 1975 through 1978. Nearly a third of it came in 1978 alone, with the death of Tito.

Thematic Content Table 3.8 showed that Eastern Europe and the Soviet Union together had the lowest proportion of crisis coverage among all major regions of the world. However, Table 3.12 makes it clear that the USSR alone accounts for that pattern. Each of the other four leading news nations of Eastern Europe show a proportion of crisis themes higher than the overall world average.

Latin America
Nearly all of the nations in Latin America were referred to at least once in the sampled news broadcasts. Table 3.13 includes the 10 most frequently mentioned nations, which together comprise 80 percent of all references to nations of the region during 1972–1981.

Among these 10 leading news nations from Latin America, the pattern of attention was slightly more dispersed than for other regions of the world. Cuba, the most frequently mentioned nation, accounted for only one-quarter of news from the region. Mexico and Panama (including the Canal Zone), each received about one-half that amount of coverage. Chile, El Salvador, Argentina and Nicaragua are clustered at a third level of coverage, each comprising from 5 to 6 percent of national references within Latin America. Finally, Brazil, Venezuela and Puerto Rico received low levels of network news attention.

Report Formats As shown in Table 3.13, more than half of the references to Brazil and Nicaragua occurred in foreign video reports. With the exception of those nations and El Salvador, all of the leading news nations of Latin America had a low proportion of references in foreign video reports, following the pattern for the entire region. In other words, for seven leading news nations of the region, the networks either could not, as in the case of Cuba early in the decade, or did not send their own correspondents to cover events directly.

The lack of permanent network bureaus in Latin America affects not only coverage of the region's news leaders, but also coverage of news in other nations when it occurs. For example, when a military coup took

Table 3.13
Coverage Given the News Leaders of Latin America, 1972–1981

Nation	N = (references)	% of References[a]	Story Format[b]			Crisis Themes[c]	
			Anchor %	Domestic %	Foreign %	%	N
1. Cuba	223	24.5	30.0	47.5	22.4	25.3	75
2. Mexico	117	12.9	41.9	35.0	23.1	26.7	45
3. Panama and Canal Zone	104	11.4	26.0	47.1	26.9	11.4	44
4. Chile	57	6.3	47.4	31.6	21.1	39.1	23
5. Argentina	51	5.6	58.8	13.7	27.5	52.6	19
6. El Salvador	48	5.3	20.8	45.8	33.3	55.6	18
7. Nicaragua	47	5.2	23.4	23.4	53.2	62.5	16
8. Venezuela	33	3.6	39.4	39.4	21.2	22.2	9
9. Puerto Rico	23	2.5	30.4	52.2	17.4	25.0	8
10. Brazil	21	2.3	14.3	33.3	52.4	20.0	5
N = (references)	909						

[a] The base for these percentages is 909, the total number of references to nations of Latin America, including some not in the above table.
[b] The base number for all story format percentages is the number of story references for each nation, found in the first column of the table. Percentages for the three story formats may not sum to 100 percent due to rounding error.
[c] The proportion of crisis themes for each nation is based on data from the "CBS Evening News" only, 1972–1981.

place in Guatemala on March 23, 1982 coverage of the event was very similar on each of the three networks. On the evening of March 23, the coup was the lead story on two of the three networks, but none of them had a correspondent on the spot to cover developments. By the next day, all three networks had flown correspondents to Guatemala City, and each carried a report from the capital city.

Four Latin American nations received a relatively high proportion of references in domestic video reports, indicating in each case, some direct concern or involvement on the part of the U.S. government. They included the territory of Puerto Rico, Cuba, Panama (including the Canal Zone), and El Salvador. With the exception of Nicaragua and Argentina, the remaining Latin American news leaders were well above the overall network average for this reporting format. As indicated earlier, this pattern reflects U.S. political and economic interests in the region and the tendency of the television networks to report and view Latin America as a sphere of United States influence, much like Eastern Europe is considered to be part of the "Soviet Bloc."

Puerto Rico is a U.S. territory and what happens there is usually related to domestic issues. Cuba is always of domestic concern, almost by definition, but usually not in a crisis context. For example, as these words are being written in July 1983, the problem of airline hijackings from the U.S. to Cuba has again become a major issue in the news. In the case of Panama, nearly all of the network news coverage dealt with the negotiation of a Panama Canal Treaty and was thus tied to domestic political debate. El Salvador received a great deal of coverage in 1981 in large part because President Reagan made it a policy issue.

Argentina and Nicaragua, which were not referred to as frequently in domestic video reports, can be explained differently. Argentina had internal problems and civil war but it was not directly important to the United States because it was mainly a military takeover and crackdown on leftists which never really threatened the *status quo.* Nicaragua was reported by the U.S. networks as a civil war against an unpopular dictator. The U.S. didn't take the war seriously and therefore didn't have much leverage in negotiating once it became clear that Somoza would go.

The networks relied most heavily on the wire services for coverage of Argentina, Chile and Mexico. The proportion of a given nation's coverage comprised by anchor reports may depend on several factors, including the nature of events being reported and the availability of video reports, either foreign or domestic. In the case of both Argentina and Chile, the bulk of news coverage came during the early part of the 1972–1981 decade, when satellite transmission costs were relatively high and when the network's overall use of their own correspondents was at a much lower level. In addition, the Chilean government placed restrictions on foreign news organizations for a time following the coup and assassination of President Allende. Such factors help to explain why the networks used so much news from the wire services in covering those nations. However, the amount of news agency material used in coverage of Mexico must come as somewhat of a surprise. Despite its importance as a supplier of oil to the United States and an ever closer and more important relationship as a neighbor, the U.S. television networks used news agencies about as much in reporting Mexico as they did on the average for all other nations of the world.

Trends Over Time Several Latin American nations contributed to the overall trend toward more network coverage of the region. Foremost among them are Cuba and Mexico, whose coverage increased gradually over the decade. Each of these nations also received some news coverage during each and every year of the decade.

A second group of nations which were not covered so consistently had peaks of coverage during the second half of the decade, contributing to the overall regional pattern. Panama, including the Canal Zone, received most of its coverage during the four years from 1977 through 1980 in connection with negotiation of the Panama Canal treaty with the U.S. El Salvador received two-thirds of its coverage in 1981, with the remaining third coming mostly during the two preceding years of military conflict in that nation. The nearby nation of Nicaragua also drew attention because of military conflict, with over 72 percent of its references coming during 1978 and 1979.

As discussed earlier, Chile and Argentina had a contrasting pattern of coverage, with most of their attention coming during the early part of the decade. In the case of Chile, network television coverage came with the overthrow of Allende in 1973 and subsequent military repression. In Argentina it came with the death of Peron, the ouster of Mrs. Peron, and the military repression of leftists that followed in the middle of the decade. The remaining three news leaders, Venezuela, Puerto Rico, and Brazil show no discernible trend in coverage during 1972–1981 and are characterized by minimal levels of network news attention.

Thematic Content Not surprisingly, Nicaragua and El Salvador are the two Latin American nations showing the highest proportion of crisis coverage. Network television covered war and civil unrest in both nations during the latter part of the 1972–1981 decade.

More than half of the coverage given Argentina dealt with crisis, and Chile was next highest on this measure, with nearly 40 percent of its coverage involving crisis.

Africa

As a region, Africa had proportionately more nations that did not appear at all in the sampled news broadcasts than any other part of the world. Between 40 to 50 percent of the 43 nation states in sub-Saharan Africa were not mentioned in the sampled news broadcasts.

Table 3.14 rank orders the leading news nations of Africa during the 1972–1981 period. As with the other world regions, concentration of attention on a few nations is evident with the five nations of South Africa, Zimbabwe, Angola, Uganda and Zaire accounting for 70 percent of all African coverage. Together, all 10 nations in the table comprise just over 86 percent of all African news coverage during the decade.

Report Formats Several of the leading African news nations received a relatively high proportion of their references in foreign video reports, including Zambia, Zaire, Zimbabwe, South Africa, Angola and Somalia. However, in only two of those cases did a large number of reports originate from the nation itself. They were Zimbabwe and South Africa, as shown earlier in Table 3.3.

Three of the African news leaders, Angola, Zaire, and Somalia received higher than average coverage through domestic video reports, reflecting, in each case, direct U.S. involvement or foreign policy interests relating to the nation. For example, Angola and Zaire were both reported in connection with CIA-backed forces fighting in the Angola civil war.

Most of the news leaders in Africa reflect the region-wide tendency for the networks to rely on news agencies for their coverage. Zaire, Angola and Somalia are the only real exceptions to this pattern, and in each case it would appear to be because the very limited coverage they received elicited foreign or domestic video reporting.

Trends Over Time News from Africa peaked in 1976, and continued at higher than normal levels for the next two years, as a result of events in several nations. Heavy coverage of the war in Angola began in 1976, with the intervention of Cuban troops and public discussion of CIA-backed soldiers operating out of Zambia and Zaire. For a period of time, the Angola fighting became a major issue in the U.S. Congress, which questioned the U.S. role in the conflict.

In mid-1976, an Air France plane carrying a large number of Israeli citizens was hijacked and flown to Entebbe, Uganda. The hijacking and the dramatic rescue of the hostages by Israeli commandos received intense coverage by the U.S. television networks.

The year 1976 also marked the first of four years in which the networks devoted attention to the internal conflict in Zimbabwe. Coverage of the fighting there and the transition to majority rule was heaviest in 1976 and again in 1978.

Table 3.14
Coverage Given the News Leaders of Africa, 1972–1981

Nation	N = (references)[a]	% of References[a]	Story Format[b]			Crisis Themes[c]	
			Anchor %	Domestic %	Foreign %	%	N
1. South Africa	155	23.7	40.6	23.9	35.5	40.5	153
2. Zimbabwe	136	20.8	44.9	14.0	41.2	29.4	136
3. Angola	61	9.3	31.1	34.4	34.4	49.1	55
4. Uganda	63	9.6	65.1	4.8	30.2	31.5	54
5. Zaire	42	6.4	23.8	33.3	42.9	43.9	41
6. Ethiopia	29	4.4	58.6	10.3	31.0	50.0	24
7. Somalia	24	3.7	29.2	33.3	37.5	54.5	22
8. Zambia	21	3.2	38.1	4.8	57.1	66.7	21
9. Tanzania	19	2.9	52.6	21.1	26.3	22.2	18
10. Kenya	16	2.4	56.3	18.8	25.0	6.7	15
N = (references)	655						

[a] The base for these percentages is 655, the total number of references to nations of Africa south of the Sahara, including those not named in the above table.

[b] The base number for story format percentages is the number of story references for each nation, found in the first column of the table. Percentages for the three story formats may not sum to 100 percent due to rounding error.

[c] The proportion of crisis themes for each nation is based on data from the "CBS Evening News" only, 1972–1981.

South Africa, the most frequently mentioned nation in Africa, received some attention in connection with the crises in Angola and Zimbabwe. However, much of its coverage related to other topics such as ethnic or racial discrimination, majority rule, the status of Namibia, and Davis Cup tennis matches. During the Carter administration, United Nations ambassador Andrew Young frequently appeared on network television as an outspoken critic of the South African government.

Thematic Content For the most part, the news leaders of Africa reflect the regional pattern of heavy emphasis on crisis news. The only real exceptions are Tanzania and Kenya, which received minimal levels of coverage over the decade. The proportion of crisis reporting from Zimbabwe and Uganda is low because there was considerable emphasis on negotiations and diplomacy following the actual crisis reporting in each nation. In Zimbabwe, the transition to majority rule was at times political and at other times a military struggle. Following the Entebbe incident, the networks paid greater attention to Uganda, up to the point of Idi Amin's overthrow as leader of that nation.

GEOGRAPHICAL PATTERNS IN NETWORK NEWS COVERAGE

The analysis presented in this chapter reveals three general characteristics of network television's news geography during the 1972–1981 period. They are (a) hierarchy or concentration, (b) inter- and intraregional differences in coverage, and (c) changes in coverage over time. These three characteristics help to summarize the overall pattern of network television's news geography.

Hierarchy and Concentration

Network television news is highly concentrated, with a few nations of the world receiving the bulk of the coverage at a given point in time and most nations being ignored. This finding holds true whether the analysis is at a global or regional level. Such a descriptive finding, per se, is hardly surprising. In fact, to some extent it appears to reflect the very definition of news. It would be unrealistic to expect all nations of the world, or even all nations within a particular region, to be equally newsworthy at all times.

A more important issue raised by this finding concerns the degree of concentration in network television's international affairs coverage. It is significant that network television's direct, visual coverage from overseas is much more concentrated than its total coverage, based on all national references in the news. Although this study has made no such comparisons, the question of whether television's international news is more concentrated than that of other media is also a very important one.

Differences in Coverage

During the years from 1972 through 1981 there were a number of differences in the extent of coverage given to nations and regions. Nations like the USSR, Israel, Vietnam, Iran, Britain, Egypt, France, the People's Republic of China, Lebanon, West Germany, and Japan were news leaders of the world. At a regional level, Western Europe, the Middle East, and Asia were clear leaders, and Latin America and Africa were blind spots, relatively speaking, on the network news. In addition to differences in the amount of attention given to major world regions, certain of the regions differed in the degree of concentration of coverage within the region. Africa, for example, showed a much more concentrated pattern of coverage than Western Europe or the Middle East.

Nations and regions also differed on the proportion of coverage accounted for by each major story format, anchor reports, domestic video reports, and foreign video reports. For example, Western Europe and the Middle East receive a relatively high portion of their coverage in the form of foreign video reports, which show direct visual newsgathering by the networks own correspondents. On the other hand, Eastern Europe, Africa and Latin America receive a relatively low proportion of their coverage in the foreign video format. The networks rely more heavily on news agencies for coverage of Africa than for news of other regions. Such patterns illustrate the relationship between international news content on television and the newsgathering process that produces the content.

Another difference among nations and regions in network news coverage is the relative proportion of crisis coverage devoted to each. At the regional level, the contrast between Africa and Eastern Europe is instructive. Africa receives the highest proportion of crisis coverage, more than double the proportion given Eastern Europe and the USSR. Does this difference reflect only the incidence of crises in these two regions, or does it also say something about priorities and constraints in the newsgathering process?

Changes Over Time

The amount and nature of news coverage changed over the 10 years at both national and regional levels. The Indochina nations of Vietnam, Kampuchea, Laos and Thailand provide a good example. Early in the decade they received very high levels of coverage, and a high proportion of it was crisis reporting of the Indochina war. During the last half of the decade, their levels of coverage dropped dramatically, as did the proportion of crisis content in such news.

In summary, the changes in network television's international news coverage between 1972 and 1981 appear to fit the general pattern that has been observed in previous studies of print media. If anything, the pattern is more pronounced for television because of its appetite for hard news with visually exciting elements. There are some nations, and some large regions, which receive only relatively brief but intense periods of coverage when a major crisis occurs. Over the long term, these represent disruptions in the more stable pattern of more or less continual attention to certain nations and regions. During 1972–1981, disruptions occurred in such nations as Angola, Zimbabwe, Zaire, Chile, and Pakistan. Nations like the USSR, Britain, France, Japan, and West Germany exemplified more stable patterns of coverage over the long term.

Finally, there are a group of nations which seldom, if ever, appear on network television news. It includes such countries as Bhutan, Burma, Mongolia, Sierra Leone, Senegal, Rwanda, Chad, Bulgaria, Benin, Botswana, Belize, Malta, and Liechtenstein. These are the nations whose developments are not visible through network television's window on the world. Most of them are in the Southern Hemisphere and comprise part of the Third World. The following chapter will more explicitly compare coverage of the Third World with coverage of developed and socialist nations.

Chapter 4

Coverage of Developed, Developing, and Socialist Nations

How well do major Western news media cover events in developing nations, or in nations with different political and economic systems? For reasons briefly touched on in Chapter 1, this question is part of the ongoing international discussions about the shape of a "New World Communication Order." It is a politically volatile issue because it involves not only the amount of news coverage given to different categories of nations, but also the question of values or selection criteria that constitute the prevailing definition of news in the Western media. It is also an issue of great concern in all categories of nations, large and small, developed and developing, capitalist and socialist.

The purpose of this chapter is to compare network television's coverage of developed nations with its coverage of developing and socialist bloc nations during the 1972–1981 decade. The principal focus is on the coverage of developed and developing nations. In a broad sense, the chapter addresses the question of how well television conveys to the American public and U.S. government policymakers an accurate picture of North–South relations and of important social change and development issues in the Third World.[1] A secondary focus is on coverage of socialist nations in comparison with the other two categories of countries.

The first section of this chapter examines the issue as a political and policy concern. The second section reviews the nature of existing research. Then the content data from U.S. network television broadcasts during 1972–1981 are used to compare coverage of developed and developing nations on this particular medium.

THE PROBLEM OF THIRD WORLD COVERAGE IN MAJOR WESTERN MEDIA

The International Commission for the Study of Communication Problems (the MacBride Commission) noted several quantitative imbalances in the circulation of news, including

---

[1] The terms "Third World" and "developing" are used interchangeably.

1. the flow of news between developed and developing nations,
2. the flow of news among developing nations themselves, and
3. the flow of news between nations having different socioeconomic and political systems (International Commission for the Study of Communication Problems, 1980, p. 36).

In addition to such quantitative imbalances in the flow of news, the MacBride Commission report also summarized several qualitative imbalances, including those (a) "between political news and news concerning the social, economic and cultural life of countries battling with the ills of underdevelopment" and (b) between "good" news and "bad" news (1980, p. 36). The first of these qualitative imbalances calls attention to the issue of development news, or the lack of it, in Western media. The processes of social change and development are often gradual, long-term and nonspectacular events which do not conform to prevailing Western news values, particularly for a visual medium like television. During the 1970's, policymakers and leaders from a number of developing nations repeatedly called attention to the lack of serious, sustained development news in the major Western media, a shortcoming that has been at least partly acknowledged by media professionals in the West (Rosenblum, 1979, p. 203–213). The problem was articulated as follows by Narinder Aggarwala, a prominent Third World spokesman,

> The general public in the West gets to look at world events through the prism of their media. In the developing countries also, people come to see their own world through the eyes of the Western media. Quite often, the media transmit single-dimensional, fractured images perceived by viewers as reflections of the whole. Partly this is due to the nature of the craft and partly due to an overemphasis on spot or action news in international news dissemination systems. Western media leaders insist that noncrisis news is of little interest to the general public. But news is what happens, and the most important thing happening in the Third World today is the struggle for economic and social change. It is imperative for the survival of a free press that journalists and media leaders find ways to cover development news interestingly and adequately. (Aggarwala, 1981, p. xix)

The second qualitative imbalance, between "good" and "bad" news, refers to the often observed tendency of the news media to focus on failure, catastrophe, or the negative side of events in general. Both the quantitative, and to a more limited extent, the qualitative imbalances in the flow of news will be explored later in this chapter as they are reflected in network television's coverage of international affairs.

Although the issue of how well developing nations are covered in Western media, including U.S. television, is international in scope, it has important policy implications in the United States. In the words of Otis Chandler, publisher of the *Los Angeles Times,*

> the problem of obtaining pertinent information about Third World nations has assumed dimensions of such national importance that it should have the attention of everyone. This is because the energy crisis plus recognition that accommodation must be reached with the Third World have impelled the United States to formulate new policies toward the non-aligned areas of the world.
>
> Thus, the question arises: How can the government formulate and the public pass judgement on a sensible policy toward a country about which we know little or nothing beyond information supplied by the government's own agencies? (Chandler, 1977, p. S9804)

The question posed by Chandler concerning the policy implications of international news coverage by the U.S. media is an important one and a primary impetus behind the empirical analysis which follows. International news content on network television is important because of its presumed effects. The policy implications of such coverage during the 1970s will be addressed in more detail in Chapter 6.

EXISTING RESEARCH

Over the years, many books, monographs, and scholarly articles have been published on the topic of international news flow and related policy issues. In recent years the volume of such research has increased, in response to heightened international concerns about communication issues. A large portion of the research on news flow touches in some manner the question of how developed nations compare with developing nations in media coverage. For present purposes, no attempt is made to review that literature to any great extent. Instead, it is categorized according to three major approaches to the question.

Empirical studies that discuss or compare coverage of developed and developing nations generally fall into one of the following three categories: general news flow studies, comparative studies, and explanatory studies. With few exceptions, all such studies deal with print media, including newspapers, wire services and news magazines.

General News Flow Studies

A number of studies, conducted on major media in several different nations, examine international news coverage in general, or coverage of a particular region of the world (Golding and Elliott, 1979; Harris, 1974; Schramm, 1964; Gerbner and Marvanyi, 1977). Such studies do not make explicit, quantitative comparisons of coverage given developed and developing nations. Nor do they usually include a clearly stated conceptualization and operationalization of what is meant by "developed" and "developing." However, they often contain statements or general conclusions about the amount and nature of coverage given to these categories of nations. For example, Wilbur Schramm, after reviewing a number of such studies, concluded that "the flow of news among nations is thin. . . it is imbalanced, with heavy coverage of a few highly developed countries and light coverage of many less-developed ones" (Schramm, 1964, p. 65).

Comparative Studies

A second category of studies, much smaller in number, uses descriptive content data to compare coverage of developed and developing nations, usually treating development as a dichotomous variable (Hester, 1971; Larson, 1978, 1979; Weaver and Wilhoit, 1981; Stevenson and Cole, 1980; Pasadeos, 1982). Such studies have generally shown that developing nations receive more total coverage than developed nations. However, on a per-country basis or in relation to the population of the nations involved, the finding usually reverses, with developed nations receiving more attention in the news than developing nations. The existing studies of print media also provide some evidence that coverage from the Third World tends to contain a higher proportion of crisis news than does reporting from developed nations.

Explanatory Studies

A third category of studies bearing on the question of how developed and developing nations are covered by major news media consists of research which is explanatory in nature. Such studies treat development as a variable which, along with other "extra-media" data, may be used to predict or explain coverage in different news media (Rosengren, 1970). For example, Semmel (1977) explored the pattern of international news coverage in the U.S. elite press in an effort to explain news coverage in terms of the economic, political and cultural position of other nations relative to the United States. He operationalized economic distance of other nations from the U.S. in terms of Gross National Product Per Capita. Charles, Shore and Todd (1979) used international trade and telecommunications traffic in an effort to explain *New York Times* coverage of Equatorial and Lower Africa. Overall, however, the existing research literature contains very few studies that use measures or correlates of development in an effort to explain or predict coverage.

This chapter's analysis of U.S. network television content is descriptive and comparative in nature, placing it in the second of the three categories of research described above. Before making the comparisons, one aspect of the procedure requires a brief explanation.

THE CATEGORIZATION OF NATIONS FOR THIS COMPARISON

The comparisons in this chapter are all based on content data gathered according to the procedures described in Chapter 2. However, this chapter entails one additional procedural problem: the question of how to place the nations of the world in categories such as developed, developing or socialist bloc.

As other researchers have noted (Hester, 1971; Weaver and Wilhoit, 1981), the categorization of nations is a difficult and somewhat arbitrary task. However, in order to compare coverage of developing nations with that of other groups of nations, each nation of the world must be placed in a category according to some meaningful criteria. For the purposes of all comparisons in this chapter, data concerning coverage of dependent territories, which accounts for only about 3 percent of network international news, principally from Northern Ireland and Hong Kong, are ignored.

The categorization of nations in this chapter is based primarily on the one used by the United Nations in reporting international trade data for its annual *Statistical Yearbook* (1978). Nations are classified as either developed market economies, developing market economies, or centrally planned economies. Appendix D contains a complete listing of the nations and territories in each category.

Conceptually, the categories of developed and developing market economies correspond quite closely to developed and developing nations as they are usually referred to in discussions of news flow imbalances. However, the categorization does differ from one based exclusively on Gross National Product Per Capita. Using only a measure of GNP Per Capita as the criterion for development, a small number of island nations and oil-producing nations would be considered developed, rather than developing nations.

COVERAGE OF DEVELOPED, DEVELOPING, AND SOCIALIST NATIONS

The sample of network news content representing the 1972–1981 period allows a number of comparisons concerning the amount and nature of coverage given developed, developing and socialist nations. Each of the following research questions will be addressed through analysis of the content data file.

1. Do developing nations receive less coverage than developed nations or socialist nations on network television?
2. Are developing nations more likely to be mentioned in relation to developed nations than vice versa?
3. Are developing nations as likely to be the focus of a story themselves as are developed nations?
4. Do the networks use the three major reporting formats in the same proportions in covering developed, developing and socialist nations?
5. Does network television coverage of developing nations contain a higher proportion of crisis content than coverage of developed or socialist nations?
6. Are there changes over time in the reporting formats, crisis content, or amount of network coverage of developed, developing, and socialist nations?

The following sections of this chapter address each of the above questions in turn, using quantitative data to support findings and conclusions whenever possible.

Quantity of Coverage

As with other findings of this study, the amount of coverage given to developed, developing and socialist nations may be quantified in three different ways. First, the number of sampled news stories which mention each category of nations may be counted. Since one, two, or even all three of the categories of nations may be mentioned in a single news story, this method of quantification results in percentages that total to more than 100 percent. Second, the findings may be quantified in terms of total references to nations, remembering that only one reference for each nation was coded per news story. This approach allows a computation of the proportion of coverage accounted for by each category of nations, with percentages totaling 100 percent. Third, for foreign video reports only, the number of reports originating in each category of nations may be quantified and compared. This final method is especially important because it measures most directly the level of money and resources committed by the networks to covering foreign news in that category of nations.

Table 4.1 shows the proportion of sampled stories which mention each category of nations or combination of categories. The table includes data only for nation states, ignoring mentions of the United States and dependent territories. The seven categories in the table represent all possible combinations of these three categories of nations, if the U.S. is ignored. Fourteen categories could be constructed if mention of the U.S. is added as a fourth category. The data presented in Table 4.1 and subsequent tables in this chapter are based on data about 28 developed nations, 13 socialist nations, and 132 developing nations.

Table 4.1
Network Television International News Stories (percentage) by Category of Nations Mentioned, 1972-1981

Nations Mentioned	ABC	CBS	NBC	All Networks
1. Only Developed Nations [a]	23.3	24.1	25.0	24.1
2. Only Developing Nations	30.9	30.3	31.1	30.8
3. Only Socialist Bloc Nations	15.0	14.0	14.1	14.4
4. Developed and Developing Nations	15.2	16.0	13.6	15.0
5. Developed and Socialist Nations	4.3	4.4	4.7	4.5
6. Developing and Socialist Nations	7.6	7.4	7.6	7.6
7. Developed, Developing and Socialist Nations	3.6	3.7	4.0	3.8
N = (stories)	2302	2326	2192	6820

Raw chi-square = 7.68627   12 DF   Nonsignificant

[a] "Developed" nations refer to all those other than the United States. Mentions of the U.S. were ignored in constructing the table.

The data in Table 4.1 provide part of the answer to the first research question posed above. In terms of the proportion of news stories which mention each category of nations, developing nations received more news coverage than developed nations. Altogether, developing nations were mentioned in over 57 percent of the sampled news items (categories 2, 4, 6 and 7 from the table mention developing nations), while developed nations were mentioned in over 47 percent of sampled stories (categories 1, 4, 5, and 7 from the table). Socialist nations, which comprise the rest of the countries in the world, were referred to in approximately 30 percent of the sample news items. The chi-square value computed for this table indicates that there are no statistically significant differences among the three networks in their attention to the seven categories of nations.

Quantification of the data in terms of the proportion of total references accounted for by each category of nations results in the same pattern. Developing nations account for approximately 42 percent of all national references in the data file, compared with about 35 percent for developed nations and 22 percent for socialist nations.

For foreign video reports only, it is possible to compare the number of reports originating in developed, developing and socialist nations, respectively. The origin for such reports is important both because of the resources required to send or station correspondents overseas and because foreign video reports are arguably the most visually appealing elements of network foreign news coverage. Table 4.2 presents data on the origin of foreign video reports on ABC and CBS for the entire 10-year period and on NBC for the 6 years from 1976 through 1981. It shows that nearly half of all foreign video reports broadcast by the networks during 1972–1981 originated in the developed nations of the world. Even though developing nations outnumber developed nations by nearly a 5 to 1 ratio, fewer foreign video reports originated in the developing nations, comprising about 39 percent of such reports. The socialist nations accounted for approximately 10 percent of foreign video reports, and the remaining three percent originated in dependent territories. The chi-square statistic computed on the cross-tabulation in this table shows that there is no statistically significant difference among the three networks in their direct gathering of visual news from developed, developing, and socialist nations or dependent territories. In summary, Table 4.2 provides a different answer to the initial research question than did Table 4.1, based only on references to nations in the news. *Based on the origin of foreign video reports, developing nations do receive measurably less coverage than developed nations.*

Table 4.2
Origin of Foreign Video Reports (percentage) by Network
and By Category of Nation

Origin	ABC, 1972–1981	CBS, 1972–1981	NBC, 1976–1981
1. Developed Nations	46.4	47.7	47.1
2. Developing Nations	39.2	39.7	38.0
3. Socialist Nations	11.2	9.1	11.4
4. Dependent Territories	3.3	3.5	3.5
N = (stories)	886	711	463

Chi-Square = 2.43220   6 DF   Nonsignificant

Note: Column percentages may not sum to 100 percent due to rounding error.

Nations Mentioned in the News

Two of the research questions stated at the outset of this chapter deal with the particular nations mentioned in news coverage and the number of nations mentioned. First, do developing nations appear most often on network television in news stories that involve developed nations? Table 4.1 provides part of the answer to this question. Ignoring mentions of the

United States, well over half of those news stories that refer to developing nations mention only other developing nations. However, the United States is mentioned in approximately 64 percent of all news stories referring to developing nations. If it is considered along with other developed nations, about 78 percent of all news about the Third World also contains references to developed nations.

A second research question is whether developing countries appear less often than developed nations in television news stories that mention only a single nation. The assumption underlying this research question is that such stories represent what Golding and Elliott (1979) referred to as "real foreign news" (p. 156). That is, they involve events which are considered newsworthy in their own right, not only because of the involvement of the United States or some other developed nation. Table 4.3 cross-tabulates the first three categories of nations that can appear in news stories by the number of nations mentioned, expressed as a dichotomous variable (one nation or more than one being mentioned). It indicates clearly that developing nations are less likely than either developed or socialist nations to be mentioned alone in a network television news story, apart from references to any other nations. The chi-square value reported for the table indicates that these differences are statistically significant.

Table 4.3
International Stories that Mention a Single Nation or More Than One,
by Category of Nations Mentioned (percentage), All Networks

Category of Nations Mentioned	Number of Nations Mentioned			
	One	More Than One	N = (stories)	%
1. Developed Nations Only	44.4	55.6	1645	100.0
2. Developing Nations Only	26.0	74.0	2098	100.0
3. Socialist Nations Only	24.9	75.1	979	100.0
Average for Above Categories	32.2	67.8	4722	100.0
Chi-Square = 172.77012	2 DF	Significance = .000		

Reporting Formats

The preceding findings show that foreign video reports which may mention developing nations often originate in developed nations, perhaps due to the presence of a network bureau or better air connections for traveling correspondents. A more general question is whether the networks use the three major report formats in the same proportion for coverage of developed, developing and socialist nations. As reported earlier, all network news, on the average, consists of 41 percent anchor reports, 26 percent domestic video reports, and 33 percent foreign video reports. Table 4.4 shows the proportion of references to developed, developing and socialist nations which occurred in each of the three formats.

Table 4.4
Report of Format of International News Stories by Category of Nations Mentioned (percentage) All Networks, 1972-1981

Nations Mentioned	Anchor	Domestic	Foreign	N = (stories)
1. Only Developed Nations	43.5	18.1	38.5	1645
2. Only Developing Nations	45.7	29.7	24.5	2098
3. Only Socialist Bloc Nations	47.1	31.2	21.8	979
4. Developed and Developing Nations	33.6	21.8	44.6	1022
5. Developed and Socialist Nations	41.1	20.4	38.5	304
6. Developing and Socialist Nations	31.7	34.6	33.8	515
7. Developed, Developing and Socialist Nations	22.2	33.5	44.4	257
All Stories	41.4	26.0	32.6	6820

Chi-Square = 314.64247   12 DF   Significance = .000

Note: Some row percentages may not sum to 100 percent because of rounding error.

The data from Table 4.4 indicate that the three networks depend most heavily on the news agencies for coverage of developing and socialist bloc nations, but not by a large margin. The major differences in coverage of the three categories of nations appear in domestic and foreign video reports. Both developing and socialist nations show a higher proportion of references in domestic video reports than do developed nations. In the case of socialist nations, this may largely reflect the salience of the nations like the Soviet Union, China and, early in the decade, Vietnam on the Washington, D.C. policy agenda. When developing nations do appear in the news, it is more likely to be in connection with some direct U.S. policy interest than when developed nations appear in the news. Table 4.4 also shows that developed nations are more likely than either developing or socialist nations to be mentioned in a foreign video report. This tendency certainly reflects the presence or proximity of network bureaus and correspondents and may also reflect historic political and economic ties and the news values of U.S. media. The preceding differences are statistically significant, as indicated by the chi-square value reported with Table 4.4.

The Question of Crisis Coverage

Major Western news media have repeatedly been charged with a tendency to cover Third World nations only in times of crisis or major conflict. The charge may be especially applicable to television because of its appetite for visually exciting and captivating material (Batscha, 1975; Epstein, 1974; Rosenblum, 1979). Using the available data on thematic content of network television's international news during 1972–1981 it is possible to answer the question of whether news about developing nations contained a higher proportion of crisis stories than news from developed or socialist nations. As with most content-based studies, this approach leaves unanswered the question of what events or crises actually occurred in developed versus developing nations during 1972–1981. However, it does address an important dimension of the news coverage itself and allows some inferences to be made about the question of possible overemphasis on crises in coverage of the Third World.

The content data do show that network coverage of developing nations contains a much higher proportion of crisis content than its coverage of developed and socialist nations. Approximately 32 percent of all stories that mentioned developing nations dealt with crisis themes, as compared with 26 percent for developed and 21 percent for socialist nations. Table 4.5 shows that 36 percent of stories that mention only developing nations deal with crisis, a higher proportion than for any of the other categories of stories according to nations mentioned.

The data concerning national origin of foreign video reports provide an even more striking picture of network news crisis reporting. Nearly half (47.5 percent) of all stories reported directly from developing nations by network television correspondents deal with crisis, compared with only 32.1 percent of such reports that originate from developed nations and approximately 30 percent of foreign video reports from socialist nations. As noted in Chapter 3, the proportion of crisis themes from socialist nations would be higher if the USSR did not account for such a high proportion of coverage for that group of countries.

One plausible explanation for the higher proportion of crisis in news from the Third World is the nature of events in those nations during the time period of this study. However, data discussed in the preceding chapter make it clear that the overall pattern of crisis reporting from the Third World includes a diverse group of nations, some of which were involved in protracted war and others in which the crises were shorter and of a different nature. In any event, it is not so much the crisis news itself as the lack of noncrisis, development reporting that disturbs Third World critics of Western media, including television.

Changes in Coverage Over Time

An important question about all of the comparisons made thus far in this chapter is whether there were changes in the patterns of coverage over the 10-year time span of this study. Of particular interest are changes in the amount of coverage, reporting formats, and thematic content of news about developed, developing and socialist nations.

Table 4.5
Proportion (percentage) of Crisis Stories by Category of Nations Mentioned and Origin of Foreign Video Reports, All Networks

Category of Nations	% Crisis Themes Based on References		% Crisis Themes Based on Origin of Foreign Video Reports	
		Base N		Base N
1. Only Developed Nations	27.5	1278	32.1	872
2. Only Developing Nations	36.0	1574	47.5	693
3. Only Socialist Nations	16.1	746	29.9	194
4. Developed and Developing Nations	26.0	796		
5. Developed and Socialist Nations	17.8	230		
6. Developing and Socialist Nations	31.7	357		
7. Developed, Developing and Socialist Nations	21.6	153		
All Nations	27.9	5134	37.9	1759

Analysis of Variance Tables

	Sum of Squares	DF	Mean Square
Between Groups	24.55	6	4.0917
Within Groups	1008.47	5127	.1967
Total	1033.02	5133	
$F = 20.8017$	Significance $= .00$		

	Sum of Squares	DF	Mean Square
Between Groups	10.52	2	5.2594
Within Groups	403.56	1756	.2298
Total	414.08	1758	
$F = 22.8852$	Significance $= .00$		

Changes in Amount of Coverage

Changes in the amount of coverage given the three categories of nations may be assessed both in terms of the proportion of news stories that mention each category and in terms of the proportion of foreign video reports that originate from each category of nations. In both cases, t-tests may be used to compare the proportion of stories involving each category of nations during the first and second half of the 1972–1981 decade. Table 4.6 presents the results of t-tests conducted for that purpose.

The first set of t-tests reported in Table 4.6 shows that there were no statistically significant changes in the proportion of stories referring to developed or developing nations during the first and second 5-year period of the 1972–1981 decade. However, the proportion of all sampled stories that referred to socialist nations showed a decrease from 31 percent during the first half of the decade to 28 percent during the second five years, a change which is statistically significant at the .05 level. It is important to note that this decrease can be accounted for entirely by news references to one nation, North Vietnam, which constituted an extremely high percentage of national references during the first 5 years of the decade and a minimal percentage during the second 5 years. Except for these numerous references to North Vietnam early in the 1970s, there would be no statistically significant shift in the pattern of network news story references to developed, developing and socialist nations. According to this measure, the basic pattern of network news attention appears to be quite stable. This conclusion is based on a decade in which there were numerous peaks and valleys in coverage of individual nations and regions of the world, based on a great variety of newsworthy events.

The t-tests on data about the origin of foreign video reports on ABC and CBS show a different pattern. As indicated in Table 4.6, the proportion of foreign video reports originating in developed nations showed no statistically significant change between the two 5-year periods. However, the proportion of such reports originating in developing nations showed a decrease from the first to the second half of the decade while the proportion originating in socialist nations nearly doubled. Both of these changes were statistically significant at the .05 level, as shown in Table 4.6.

Changes in Reporting Formats

As stressed in earlier chapters, changes in the reporting formats used by the networks are of interest because of the close relationship between those formats and the newsgathering processes of television news organizations. During the 1972–1981 decade, the U.S. networks showed a trend toward greater use of domestic and foreign video reports and less use of anchor reports in their coverage of international affairs. Here the question is whether that overall pattern holds true for developed, developing and socialist nations.

Table 4.6

T-tests Comparing Mean Levels of Coverage Given Developed, Developing and Socialist Nations in 1972-1976 With Mean Levels of Coverage During 1977-1981, Based on Both References in News Stories and Origin of Foreign Video Reports

Category of Nations	Time Period	N	Mean	S.D.	T Value	DF	2-Tail Probability
Based on References[a]							
1. Socialist	1972–1976	3201	.3115	.463	2.44	6916	.015
	1977–1981	3717	.2846	.451			
2. Developed	1972–1976	3201	.4695	.499	.48	6916	.634
	1977–1981	3717	.4638	.499			
3. Developing	1972–1976	3201	.5626	.496	.01	6916	.994
	1977–1981	3717	.5626	.496			
Based on Origin of Foreign Video Reports[b]							
1. Socialist	1972–1976	653	.0735	.261	– 3.80	1506	.000
	1977–1981	855	.1345	.341			
2. Developed	1972–1976	653	.4165	.493	– .63	1506	.528
	1977–1981	855	.4327	.496			
3. Developing	1972–1976	653	.4502	.498	2.28	1506	.023
	1977–1981	855	.3918	.488			

[a] Based on data for all three networks. The mean is also the proportion of all stories that mention each category.
[b] Based on data for only ABC and CBS. The mean is also the proportion of all foreign video reports on ABC and CBS which originated in each category of nations.

106

In order to assess changes in reporting formats within each category of nations, time is treated as an independent variable and stories in each major format the dependent variable in separate regression analyses for each of the three categories of nations. The results are presented in Table 4.7.

Both developed and socialist nations showed statistically significant (at the .01 level) decreases in anchor reports. Developed nations also showed an increase in domestic video reports that was statistically significant at the .05 level, but there was no statistically significant change for socialist nations. Developing nations showed a statistically significant decrease in anchor reports and increase in domestic video reports, as shown in Table 4.7.

In order to assess the changes over time in foreign video reports, the analysis focused on data concerning the origin of such reports, available for ABC and CBS over the entire decade. The regression results are presented in Table 4.7.

Figure 4.1 shows the proportion of foreign video reports originating in each of the categories of nations in the form of a line graph. In shows visually, that depending on world events, the U.S. television networks will devote more or less attention to developed or developing nations. The regression analysis shows that for both developed and socialist nations, there was a linear trend toward more foreign video reporting while for developing nations there was a decrease in such reports. Origination of

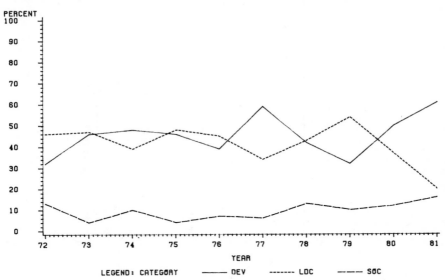

Figure 4.1
Foreign Video Coverage of Developed (DEV), Developing (LDC), and Socialist (SOC) Nations, by Year

Table 4.7
Bivariate Regression Analyses with Time as the Independent Variable[a] and Each Story Format as Dependent Variable for Developed, Developing and Socialist Nation Categories

Category of Nations	Dependent Variable	F	Significance	Simple R	R²	N = (stories)
Regressions Using National References as Dependent Variable						
1. Developed Nations	Anchor	55.618	.000	−.14884	.02215	2457
	Domestic Video	5.327	.021	.04653	.00217	2457
	Foreign Video	27.414	.000	.10509	.01104	2457
2. Developing Nations	Anchor	64.706	.000	−.12791	.01636	3892
	Domestic Video	59.289	.000	.12253	.01501	3892
	Foreign Video	.892	.345	.01514	.00023	3892
3. Socialist Nations	Anchor	20.177	.000	−.09865	.00973	2055
	Domestic Video	2.355	.125	.03385	.00115	2055
	Foreign Video	10.397	.001	.07098	.00504	2055
Regressions Using Origin of Foreign Video Reports as Dependent Variable						
1. Developed Nations	Foreign Video	12.060	.001	.08812	.00777	1543
2. Developing Nations	Foreign Video	31.139	.000	−.14074	.01981	1543
3. Socialist Nations	Foreign Video	10.300	.001	.08149	.00664	1543

[a] The independent variable, time, takes a value from 1 through 10, representing each of the years in the 1972–1981 decade.

foreign video reports from developing nations decreased noticeably in 1980 and 1981, the final two years of the decade under study.

Changes in Crisis Content of the News

Regression analysis was also used to examine changes in the proportion of crisis reporting in stories that refer to developed, developing or, socialist nations during the 1972–1981 period. Since complete thematic data for the entire decade are available only for CBS News, the analysis reported in Table 4.8 is based only on that network. However, there is no reason to believe that inclusion of data for ABC and NBC would have any effect on the results of the analysis. A comparison of crisis themes across networks for the 1976–1981 period shows no statistically significant differences in the proportion of international news stories with crisis-related themes.

As Table 4.8 indicates, developed and socialist nations showed no statistically significant increase or decrease in the proportion of crisis news referring to them over the 1972–1981 decade. On the other hand, developing nations exhibited a linear trend toward less crisis news which was statistically significant. Such a change reflects heavy network attention to several major wars or crises, such as those in Angola, Cyprus, and Vietnam during the first half of the 1972–1981 decade.

NETWORK TELEVISION COVERAGE OF THREE WORLDS

The findings presented in this chapter show how network television covered three worlds during the decade from 1972 through 1981. They are the Third World, made up of developing nations; the world comprised of the developed or industrialized nations; and finally the world of socialist nations. Several major patterns emerge from the specific findings.

First, although the absolute number of television news references to Third World nations exceeds mentions of developed or socialist countries, attention to the Third World is nowhere near in proportion to the number of developing versus developed nations. The present analysis is based on data about 174 nation states. Sixteen percent of them are considered developed, and they represent 14 percent of the population in this group of nations, yet they account for about 35 percent of all national references on network news. Seventy-six percent of the nations are classified as developing, and they account for 53 percent of the population, yet they make up only about 42 percent of all national references on network television news. Socialist nations comprise nearly 8 percent of the countries in the world, and thanks to the People's Republic of China about 34 percent of the population. They are referred to in 22 percent of network television's international news stories, largely due to heavy emphasis on the USSR.

Table 4.8

Bivariate Regression Analyses with Time[a] as the Independent Variable and Proportion of Crisis Themes as the Dependent Variable for Developed, Developing, and Socialist Nations, 1972-1981, "CBS Evening News"

| | | | | | | Degrees of Freedom | |
Category of Nations	F	Significance	Simple R	R^2		Regression	Residual
1. Developed Nations	2.949	.086	−.05122	.00262		1	1121
2. Developing Nations	19.836	.000	−.12100	.01464		1	1335
3. Socialist Nations	.254	.614	−.01929	.00037		1	685

[a] The independent variable, time, takes a value of 1 through 10, with each number representing a year during the 1972–1981 decade.

The overall lack of attention to the Third World on network television can perhaps be best illustrated visually, as in Figure 4.2. It is a world map with the 50 most heavily covered nations shaded. The remaining non-shaded nations constitute the "blind spots" of network television during the 1972–1981 decade. Each of these countries appeared in less than .7 of 1 percent (.7 percent) of all sampled coverage. These areas include some developed nations, such as Australia, New Zealand and some smaller nations in Europe and Scandinavia, along with some nations in Eastern Europe. However, the map shows at a glance that the major gaps in coverage occur among the developing nations of Latin America, Africa and Asia.

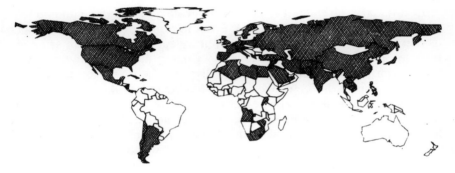

Figure 4.2
The Blind Spots of Network Television, 1972–1981
(Non-Shaded Areas)

Second, international news coverage during the 1972–1981 decade shows clearly that the networks devoted more of their own resources to coverage of developed and socialist nations than to coverage from the Third World. Nearly half of all foreign video reports broadcast by the networks during 1972–1981 originated in one of the 28 developed nations. By comparison, the number of foreign video reports originating in socialist nations increased over the decade, while those originating in developing nations showed a slight decrease. This relative lack of emphasis on direct reporting from Third World nations by network correspondents is the main finding from analysis of the three major reporting formats. In other respects, the three categories of nations followed the overall network trend of the decade toward increased use of foreign and domestic video reports and decreased reliance on the news agencies, as reflected in anchor reports.

Third, it is apparent that the U.S. networks tend to cover developments in the Third World most often when they involve the United States or other developed nations. Involvement of the "home" country, in this case the U.S., reflects a common pattern in studies of international news. However, developing nations are less likely than either developed or socialist nations to be the focus of a news report themselves. More evidence of international involvement or context appears necessary to make events from Third World nations worthy of coverage.

Fourth, television content from the decade shows that there is proportionately more crisis coverage from the Third World than from either developed or socialist nations. The emphasis on crisis is most pronounced in the case of foreign video reports that deal with events in developing nations. Nearly half of such reports involve some form of crisis. This finding corroborates observations that television news is particularly drawn toward the kind of visually exciting content that many crises provide.

In summary, although the networks devote a considerable amount of attention to Third World nations, it is not in proportion to their numbers or population. Furthermore, in terms of independent newsgathering efforts by the network organizations, developing nations receive an even lower priority. When the networks do dispatch their own correspondents to cover events in the Third World, as often as not it will be in response to an ongoing or breaking crisis. Most often, developing nations appear in news items that involve the U.S. or other developed nations, and there is less of a tendency to report news involving a single nation from the Third World than from the developed world or from socialist nations.

All of the above patterns, with the possible exception of emphasis on visual crisis news, might apply equally well to other major news media. Some, if not all, have been documented in studies of print media. Based on analysis of changes in these patterns over the 1972–1981 decade there is no evidence that the U.S. television networks are changing their approach to covering the Third World in a manner that would be responsive to current international concerns and criticisms.

Chapter 5

Some Influences on Network World News Coverage

The preceding three chapters described certain characteristics of the view provided by television's window on the world. They used quantitative data to describe the amount and nature of international affairs content during 10 years of early evening news broadcasts.

In this chapter, the focus shifts to include an examination of the window itself. Television news content is used as a dependent variable in an effort to explain some of the major influences that shape television coverage of international affairs. Specifically, the chapter focuses on the factors of satellite communication channels and location of network and U.S. news agency correspondents around the world. Both of these are important elements in network television's global news net, and are part of the chain of news communication described in Chapter 1. After a short description of the conceptual approach and research methods, the chapter presents data analysis and research findings at both bivariate and multivariate levels.

CONCEPTUAL APPROACH

In general, the conceptual approach used in this chapter is an elaboration of the theoretical approach described in Chapter 1. It suggests that there is a chain of news communication, in reality multiple chains, forming a news net which stretches from world events to viewer perceptions. Each chain involves network television and related organizations which gather and transmit international news. At various points along the chain, different influences work to expand or constrict the flow of news information. This chapter focuses on several influences which comprise part of the "technical structure" of international news communication. As Harris (1976) notes,

The international news media are engaged in translating selected world events into a saleable commodity; to present a systematic and coherent version of the events to which the news media are attracted. Which events are selected is determined in part by the technical structure of the international network of communications, by such factors as communication channels and correspondent locations. (Harris, 1976, p. 4)

The following analysis will focus on three elements in the technical structure for communication of international news for television:

1. satellite communication channels,
2. network bureau locations, and
3. location of corresondents for the two major U.S. news agencies, AP and UPI.

All three of the major U.S. television networks use satellite communication channels on a regular basis, and they often pool their transmissions through such channels. All three maintain bureaus in the major "news centers" of the world. In addition, all of the networks rely heavily on AP and UPI, along with the *New York Times* and to a lesser extent Reuters and AFP, for their basic intelligence information about what is happening in the world on a given day. Because of such similarities in the news process and because complete data are available for only two of the three networks, the following analysis focuses on ABC and CBS news during the 1972–1981 period. A preliminary analysis of NBC News for the 1977–1981 period indicates that none of the findings reported here would change with inclusion of data concerning the third network. Omission of NBC in this chapter is simply a practical problem relating to coding and structuring of the content data file over a period of several years.

In order to answer questions concerning the influence of bureaus, news agency correspondents and satellite earth stations on network coverage of international news, a more detailed discussion of the chain of news communication is needed, indicating the relative magnitude of news flow through each link or channel in the chain. Without such information it is difficult to assess the role of each channel or link in the communication infrastructure of international news flow.

As a practical matter, the flow data that would be ideal for the present analysis are extremely difficult to gather, if they are available at all. These would be scalar data on the quantity of flow through each link in the chain. However, it is possible to describe the news chain for network television in terms of either *major flows,* which account for a substantial portion of the stories, or *minor flows,* which account for a small or negligible portion of the stories. Also, there is a different pattern of flow for foreign video reports and anchor reports, the two major types of stories that originate outside the United States.

Figure 5.1 shows the chain of news communication through which anchor reports and foreign video reports flow from other nations to network anchor locations (usually New York City) for broadcast by the U.S. television networks. The major flows are indicated by solid lines and the minor flows by broken lines.

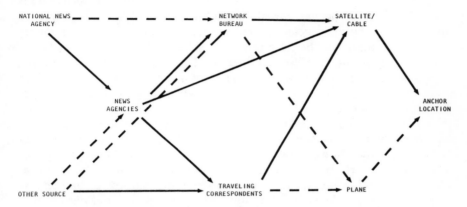

Figure 5.1
The Flows of Foreign Video and Anchor Reports From Other Countries
to Anchor Location for Broadcast on the Network News

National news services, where they exist throughout the world, are principal sources of news for the four major international news agencies, including AP and UPI (Rosenblum, 1977). Other sources within nations account for only a minor part of such news flow, as indicated in Figure 5.1.

During the 1972–1981 period, AP and UPI relied on an interconnected system of satellite and cable channels to transmit news to New York ("How News Travels," 1977). Sometimes a story may be routed circuitously through that system of interconnections, but it routinely arrives within a matter of seconds. Therefore, Figure 5.1 shows major flows from the news agencies, through satellite and cable channels, to New York. In the late 1970's and early 1980's, the relative importance and use of satellite channels continued to increase.

As indicated in Chapter 1, more than three-quarters of the anchor reports broadcast on network television news originate with the international news agencies, providing the basic outline for the day's early evening news telecast (Batscha, 1975, p. 122). The output of the news agencies is also used by network news headquarters and overseas bureau locations as a guide to gathering visual news of the day. For these reasons, only solid lines indicating major flows emanate from the news agencies in Figure 5.1.

Figure 5.1 also traces the major and minor flows of visual news, in the form of foreign video reports during the 1972–1981 period. As just indicated, the network bureaus overseas, along with network news headquarters, are heavily dependent on the news agencies for decisions on which stories to film (Batscha, 1975, p. 124). In addition, correspondents who travel from network bureaus to provide regional coverage or accompany the president or other U.S. government officials on overseas trips provide another major flow of foreign video reports to network anchor locations.

Larson (1978) formally investigated the extent to which correspondents from CBS News overseas bureaus tended to cover adjacent countries or those within the same geographical region. The extent of such regional coverage is important because of the hypothesized influence of network bureau locations in the following analysis. The acceptability of attributing news coverage of a nation to the presence of a network bureau depends in part on the relative magnitude of regional versus within-country coverage by the correspondents of that bureau. During the 1972–1976 period, the 13 nations with CBS Bureaus accounted for approximately 60 percent of all foreign video reports broadcast on the "CBS Evening News." The majority of the remaining 40 percent of foreign video reports was accounted for by two factors: (a) coverage by correspondents accompanying the U.S. President or other high officials on overseas trips, and (b) global deployment of correspondents which follows no apparent regional pattern.

Today, once foreign video reports are taped they are routinely transmitted to New York or London via satellite. By contrast, during the first half of the 1972–1981 decade some such reports were still shipped to New York by plane. Epstein (1974) discussed this practice in relation to coverage of the war in Vietnam. However, Sid Feders (1978), who became foreign editor of CBS News in 1974 estimated that three out of four foreign video reports broadcast on the CBS Evening News during 1972–1976 were fed by satellite. According to Feders, the major reason for using satellite feeds during that period of time was to provide "same-day" coverage of events.

Based on the preceding elaboration of the conceptual approach set forth in Chapter 1, the analysis in this chapter will be presented in two components. The first is hypothesis-testing, in which the following bivariate hypotheses will be tested.

1. Nations where UPI and AP correspondents are stationed will receive more coverage on U.S. network television than nations without such correspondents.
2. Nations with permanent network news bureaus will receive more coverage on U.S. network television than nations without bureaus.
3. Nations that possess Intelsat earth stations will receive more coverage on U.S. network television than nations without such stations.

The second component of the following analysis is multivariate in nature. It describes the overall pattern in which the three independent variables in the above hypotheses influence the content of network television news. Two additional variables, the population of each nation, and its Gross National Product Per Capita also enter into the multivariate analysis.

METHODOLOGY

As already indicated, a major difference between the analysis presented in this chapter and that of the preceding chapters is that content data become a dependent variable, while the independent variables involve data gathered from other sources. Rosengren (1970) calls such data *extramedia data*. Because of this shift comments concerning several aspects of the research methods will help to clarify the analysis and presentation of findings.

Units of Analysis

The news story remains the basic unit of analysis in this chapter. However, it is aggregated to the level of nations. Both dependent and independent variables are expressed as characteristics of nations. Data concerning population and gross national product per capita were not available for certain nations, principally small island states. This was also the case for nations of questionable political status, such as Namibia. No dependent territories are included in the analysis, with the exception of Northern Ireland, which is included with Great Britain for the purpose of this chapter. With these qualifications, the data file for the following analysis includes 145 nations and excludes no major nation state in the world.

Dependent Variables

Anchor reports and foreign video reports are the two dependent variables in the following analysis. Domestic video reports are excluded because they do not flow through the chain of news communication from other nations to the United States and are not subject to the influence of the chosen independent variables.

Both dependent variables are expressed in terms of the number of foreign video or anchor reports, respectively, for each of the 145 nations during the 1972–1981 period. However, the two dependent variables are quantified differently. Foreign video reports represent those that actually originated or were taped in each nation. Anchor reports, on the other hand, refer to the total number of reports mentioning each nation.

In their original form, both dependent variables are positively skewed, raising some possible questions concerning the use of parametric statistics that assume a normal distribution. In earlier research (Larson, 1978) non-parametric statistics were used to deal with this situation. However, para-metric statistics are used in this chapter since they are sufficiently robust that the findings are not affected. The findings were all cross-checked with nonparametric approaches if there was any question about the effect of skewness.

Independent Variables

The following analysis will explore the relationship of six independent variables on the two dependent content variables. The independent variables are:

1. *ABC and CBS Bureau Location.* During the 1972–1981 period, all of the U.S. television networks maintained permanent bureaus in nine nations. They included Great Britain, France, West Germany, Italy, the USSR, Lebanon, Israel, Egypt, and Japan. Each network also maintained a bureau in South Vietnam until the U.S. withdrawal from that country in April of 1975. These 10 bureaus comprise the locations that are used in the analysis for this chapter, since all were permanent bureaus maintained (with the exception of Vietnam) for the entire 10-year period. In addition, all networks had a bureau in the British Crown Colony of Hong Kong, which is omitted from the present analysis because it is a dependent territory. The People's Republic of China is excluded from the list because network bureaus there did not open until fall of 1981.

Finally, it is noteworthy that the U.S. television networks maintained some presence at different times in the following 11 nations: Singapore, Thailand, South Korea, India, South Africa, Kenya, Spain, Greece, Poland, Mexico, and Venezuela. However, through direct correspondence with the networks, (Chandler, 1982; Corrigan, 1977; Ehrlich, 1977; Kastelnik, 1982; Plante, 1983) it appears that most, if not all of the preceding 11 nations were covered by stringers or virtual one-man operations. For this reason, and because none of them were maintained as "full-fledged" bureaus for a significant portion of the 1972–1981 decade, they are not counted as bureaus in the following analysis. The 10 bureaus which are used in the analysis correspond with the list provided by Mosettig and Griggs (1980).

2. *Associated Press (AP) and United Press International (UPI) Presence.* The two major U.S. news agencies, AP and UPI maintained a presence in 50 to 70 nations during 1972–1981. These countries are listed by year in Appendix D.

3. *Intelsat Earth Stations.* At the beginning of the decade under study, 43 of the nations in the data file possessed Intelsat earth stations or access to a nearby earth station through terrestrial links. By the end of the de-

cade that number had increased to 110. As with the preceding two independent variables, the Intelsat variable is dichotomous and takes a value of 1 for each year that a nation possessed an earth station and a value of zero for the remaining nation-years in the data file.

4. *Population.* Population data for each nation were gathered from various issues of the *World Bank Atlas.* Population is a continuous variable, expressed in thousands of people per nation per year. For most nations in the data file, population figures were available for 8 of the 10 years in the study. 1977 data were used for 1978 and 1980 figures for 1981 for most nations.

5. *Gross National Product Per Capita.* Data on Gross National Product Per Capita are expressed in dollars and were also gathered from the *World Bank Atlas.* As with population, it is a continuous variable and data for two missing years were handled in the same manner as for the population variable. Gross National Product Per Capita is one measure of the level of development of nations.

6. *Nation Category.* A final independent variable used in the following analysis is the categorization of nations used in Chapter 4. The variable is categorical and takes a value of 1 for developed nations, 2 for developing nations, and 3 for socialist nations. Along with population and Gross National Product Per Capita, it may be viewed as a control variable used to examine whether correspondent, bureau and satellite earth stations influence television news coverage of different kinds of nations in the same way or to the same extent.

In an earlier study (Larson, 1978), the presence of national news agencies was used as an independent variable in the analysis of data from 1972 through 1976. However, it did not add much predictive power when used in conjunction with the first three independent variables listed above and therefore is omitted from the present analysis.

FINDINGS

Findings concerning the relationships among the above variables are presented first at the bivariate and then at the multivariate level. The t-test was used to test mean differences in levels of (a) anchor reports and (b) foreign video reports between nations with and without network bureaus, AP or UPI News Agency presence, and Intelsat earth stations. As the results presented in Table 5.1 indicate, mean levels of both foreign video and anchor report coverage are higher for nations with news agency presence, network bureaus and Intelsat earth stations. All of the mean differences are statistically significant at the .01 level.

Table 5.1
Comparison of Mean Levels of Television News Coverage for Nations With and Without U.S. News Agency Presence, Intelsat Earth Stations, and Network Bureaus

Report Format	Nation Group	Mean[a]	S.D.	T-Value	Two-Tail Probability	DF[b]	Mean Population (000)	Mean GNP/Capita ($)
Anchor	With Intelsat	3.73	7.96	3.67	.00	1284	42,273	2,905
	Without Intelsat	1.97	9.99				8,763	1,256
Foreign Video	With Intelsat	1.59	5.75	4.87	.00	1139		
	Without Intelsat	.46	2.73					
Anchor	With News Agency	5.90	13.50	8.37	.00	591	47,900	3,010
	Without News Agency	1.04	3.00				13,395	1,592
Foreign Video	With News Agency	2.42	7.18	7.26	.00	567		
	Without News Agency	.21	.88					
Anchor	With Bureau	21.67	24.75	7.85	.00	93	70,025	4,225
	Without Bureau	1.61	4.30				23,669	1,993
Foreign Video	With Bureau	9.73	14.06	6.39	.00	93		
	Without Bureau	.46	1.95					

[a] The mean number of reports per year, for each nation group listed in the table.

Note: The *World Bank Atlas* was the source of data on population and Gross National Product Per Capita.
[b] The degrees of freedom reflect use of the "nation/year" as case, with 1,450 cases (data for 145 nations, 10 years).

Also, Table 5.1 presents the mean levels of population and Gross National Product Per Capita for each of the groups compared in the first set of t-tests. These mean differences are also all statistically significant, indicating the importance of using population and Gross National Product Per Capita in further analysis.

In summary, the analysis to this point indicates that nations which have network bureaus, AP or UPI corresondents, and Intelsat earth stations have higher average levels of television news coverage than nations outside this technical structure of the international news communication system. Also, those nations within the structure have higher mean levels of population and Gross National Product Per Capita than nations outside the structure.

The Influence of Satellite Communication

To test the hypothesized influence of satellite communication on television news reporting during 1972–1981 a different analytical approach is possible which takes into account the time dimension of this study. The decade under examination was a period of rapid growth in the Intelsat global satellite system during which the number of nations possessing earth stations or connections to such stations nearly tripled. The purpose of this chapter is to assess the impact of that change on network television news coverage as completely as possible, given the available data. Before doing so it may be helpful to review the evidence already presented in Chapter 1 concerning the influence of satellite technology on television news reporting.

First, it was established that the U.S. television networks, from the time of the earliest communication satellites, were eager to make the fullest use of the new technology to improve their newsgathering capability (*Communications Satellites*, 1967). Producers place a high value on reports transmitted from a distance, particularly if they contain scenes of actual or potential violence. The testimony of network executives makes it clear that satellite technology plays an integral role in the network news process (Crystal, 1977).

Second, the parallel development of electronic newsgathering technology has made it technically much faster and easier to transmit visual reports from many nations around the world. The question of access to earth stations or other satellite links remains a political barrier occasionally encountered by television news teams, but the purely technological problems have been solved.

The hypothesized influence of satellite technology on television news may be most clearly analyzed in relation to foreign video reports as a dependent variable. An earlier study (Larson, 1978) suggested that satellite communication might also influence anchor reports on network television by facilitating telephone and other forms of communication by the international news agencies on which the networks so heavily rely for information that becomes anchor reports. The following analysis concentrates on the potential influence of communication satellites on foreign video reports by the ABC and CBS television networks because the influence is more direct, making it easier for the networks themselves to tape and broadcast visual news from more locations around the world.

In addition to the factors already reviewed, Table 5.2 shows that, as the Intelsat global system expanded during 1972–1981, the proportion of foreign video reports originating from nations using Intelsat also increased, with the major part of the increase occurring during the first half of the decade. Although such a trend would be required to demonstrate the influence of satellite technology on television news, it alone is not sufficient to do so. Other factors, such as each nation's population, level of development, or amount of trade with the United States, might explain both the increased news coverage and the expansion of Intelsat.

In view of possible rival explanations, a stronger method of testing the hypothesized influence of satellite communication on foreign video coverage by ABC and CBS is to divide the sample of 145 nations into three groups:

1. those that possessed Intelsat earth stations or connections to stations for the entire 1972–1981 period,
2. those that acquired earth stations or connections during the same 10 years, and
3. those that had no earth stations or other access during the same period of time.

Appendix D includes the relevant Intelsat data and its sources.

If use of satellite transmission increases the number of foreign video reports, Group Two nations, which obtained access to Intelsat during the period of the study would be expected to show more of an increase in this content dimension than they would have without the satellite connections. However, there is no randomly assigned control group that could allow an experimental comparison of the influence of satellite technology.

An alternative is to use both Groups One and Three to bracket Group Two and serve as control groups. Group Three, the nations which never had Intelsat connections, is not in itself an adequate control group. Among other possible differences, these nations tend to be smaller and less developed. Likewise, nations in Group One are not an adequate control because they are predominantly larger, more economically developed nations of the world. However, when both Groups One and Three are used, they are more likely to provide an adequate control. If satellites had no effect on network news use of foreign video reports, the use of such reports

Table 5.2

Proportion (percentage) of Nations With Intelsat Earth Stations and the Proportion of Foreign Video Reports Originating from Each on ABC and CBS, by Year (1972-1981)

	1972	1973	1974	1975	1976	1977	1978	1979	1980	1981	1972–1981
Proportion of Nations With Intelsat Earth Stations $N = 145$	29.7	35.9	38.6	44.8	51.0	55.9	62.8	67.6	72.4	75.9	53.4
Proportion of Foreign Video Reports Originating From Such Nations	44.3	60.2	54.0	73.9	90.3	94.4	87.2	92.5	90.7	85.7	80.0

from Group Two nations should increase at a rate between that of Group One and Group Three nations. On the other hand, if satellites had a positive effect on the number of foreign video reports, nations in Group Two should show a greater gain in the number of such reports over the 10 years than nations in either of the control groups. Finally, if satellites had a negative effect on use of foreign video reports, nations in Group Two should show less growth in the number of such reports than nations in the control groups.

Trends were analyzed based on data for the first two years and the last two years of the 1972–1981 period. Nations in Group Two had the lowest mean level of foreign video reports during the first 2 years of the period and showed a greater gain by the final 2 years than Group Three nations, but less gain than nations in Group One. These changes can be quantified using analysis of covariance. The 1980–1981 foreign video levels are the dependent variables; the beginning (1972–1973) mean levels of foreign video reports are covariates, and the grouping of nations according to Intelsat access is the independent variable.

Table 5.3 presents the deviation from the grand mean for each Intelsat subgroup, adjusted for the initial (1972–1973) mean levels. Also reported are the F-ratio and its significance level for the main effect, and the total N on which the analysis was done.

Table 5.3
Deviation from Grand Mean (3.26) on Final Foreign Video Report Levels
By Nation Groupings According to Intelsat Access, Adjusted
For Beginning Levels of Foreign Video Reports

National Group	Adjusted Deviation	N = (nations)
1. Intelsat for 10 Years	3.33	43
2. Acquired During 1972–1981	− .50	70
3. No Intelsat Access During 1972–1981	− 3.39	32
Total N		145 (nations)
F Value (significance level) for Intelsat Main Effect in Analysis of Covariance		2.058 (.132)

As Table 5.3 shows, Group Two nations which acquired Intelsat earth stations or access to one during 1972–1981 showed less of a gain in foreign video reports than Group One nations (those with earth stations for the entire 10 years) and more of a gain than the Group Three nations (those with no access to Intelsat during the decade). However, none of these differences is statistically significant, as shown by the lack of a significant main effect for the categorical Intelsat variable.

There are several possible explanations for the above finding. One crucial assumption of the above design is that the effects of satellite earth stations for countries in Group One are largely reflected in the number of foreign video reports originating from them in 1972–1973 and thus would not affect the increase in that group so sharply as for Group Two nations, which acquired the Intelsat earth stations. However, it is possible that satellite channels between the U.S. and nations in Group One were not used in 1972–1973 because of high transmission costs or the relatively cumbersome nature of television newsgathering equipment in comparison to today's lightweight cameras and editing equipment.

Also, it is possible that a nation's immediate political or economic importance to the United States is a more important influence on television news coverage than such long-term overall developments as the growth of a new communication infrastructure. Such an explanation would account for a great deal of foreign video coverage during the decade under study including coverage of wars in the Middle East and the 1973 oil embargo by OPEC nations, the Indochina conflict early in the decade, and conflicts in Nicaragua, El Salvador, and later in Iran. The list could go on, but the common denominator in such examples is the immediate political or economic impact that such events generate within the United States.

In summary, the analysis to this point has shown that satellite technology is related to the changes in television news coverage of international affairs during the 1970's. However, in terms of any causal inference, satellite technology would be termed by social scientists a "necessary" but not a "sufficient" cause of increased foreign video reporting.

Bivariate Correlations

Table 5.4 presents the Pearson correlation coefficients for each of the independent variables with (a) anchor report and (b) foreign video report coverage on the ABC and CBS early evening news broadcasts during 1972–1981. For both story types, the presence of a permanent network bureau correlates most strongly with extent of television news coverage. The correlation is somewhat stronger for anchor reports than for foreign video reports. The presence of AP or UPI also correlates with both dependent variables, but the correlation is not as strong. As expected, AP or UPI presence correlates more strongly with anchor reports than with foreign video reports. Finally, presence of an Intelsat link, population, and Gross National Product Per Capita also correlate with television news coverage, but the relationships are weaker than for network bureau or U.S. news agency presence.

Multiple Regression Analysis

In order to show the relative influence of each of the independent variables on television news content in a multivariate context, each independent variable was entered in a regression analysis with each of the dependent

Table 5.4
Pearson Correlation Coefficients for Independent Variables With Dependent Variables of Anchor and Foreign Video Report Coverage on ABC and CBS, 1972-1981[a]

	Anchor Report	Foreign Video Report	Population	GNP/Capita	Intelsat	Network Bureau
Foreign Video Report	.57 (.001)					
Population	.21 (.001)	.09 (.001)				
GNP Per Capita	.05 (.035)	.11 (.001)	-.05 (.018)			
Intelsat	.10 (.001)	.12 (.001)	.18 (.001)	.24 (.001)		
Network Bureau	.55 (.001)	.49 (.001)	.12 (.001)	.16 (.001)	.18 (.001)	
News Agency	.26 (.001)	.23 (.001)	.18 (.001)	.20 (.001)	.47 (.001)	.33 (.001)

$N = 1450$ (nation/years)

[a] Levels of statistical significance are given in parentheses.

variables. The independent variables were entered into the regression in two steps. The first step entered the Intelsat, network bureau and U.S. news agency variables, all of which are part of the technical structure of the international news system which is of primary theoretical interest in this analysis. The second step entered national population and Gross National Product Per Capita, two variables which are broader characteristics of the 145 nations in the analysis and which serve primarily as controls. Summary results of the regression analyses are presented in Table 5.5.

Table 5.5
Multiple Regression Analysis With Anchor Reports and
Foreign Video Reports as Dependent Variables

Dependent Variable	Step	Variable Entered	B [a]	F	Significance	R Square
Anchor Reports	1	Intelsat	− 1.118	6.19	.013	.0095
		Bureau	18.802	499.77	.000	.3012
		Agency	1.875	15.69	.000	.3105
	2	Population	.135	39.19	.000	.3299
		GNP/Capita	− .837	2.02	.155	.3309
Foreign Video Reports	1	Intelsat	− .850	.12	.729	.0148
		Bureau	8.704	360.61	.000	.2442
		Agency	.693	7.22	.007	.2485
	2	Population	.127	1.16	.281	.2490
		GNP/Capita	.395	1.52	.219	.2498

$N = 1450$ (nation/years)

[a] Unstandardized regression coefficient.

In the first step of the regression analysis, Intelsat earth stations, network bureaus, and U.S. news agencies together accounted for 31 percent of the variance in anchor reports and 24.8 percent of the variance in foreign video reports. For both of the dependent variables, network bureaus were the strongest predictors, followed by presence of a U.S. news agency. Presence of an Intelsat connection had a negative relationship with anchor reports that was statistically significant at the .01 level, but did not have a statistically significant relationship with foreign video reports as the dependent variable.

The second step of the regression analysis, in which population and Gross National Product were entered, changed the picture very little in the case of foreign video reports. Neither population nor Gross National Product Per Capita produced statistically significant changes in the amount of variance accounted for. However, in the case of anchor reports as the dependent variable, population increases the amount of variance accounted for in the regression by about 2 percent.

In summary, multiple regression analysis shows that presence of a network bureau is relatively the strongest predictor of network television's international news coverage. Presence of either UPI or AP is also a statistically significant predictor of network news coverage, but it is not nearly as powerful as permanent network bureaus.

SUMMARY

The analysis in this chapter showed that the technical structure of the international news communication system does relate to network television's coverage of international news. Three components of that structure, AP and UPI presence, permanent network bureaus, and Intelsat earth stations all correlate with international news in the form of both anchor reports and foreign video reports.

In an earlier study, Larson (1979) reported findings which were consistent with the hypothesis that satellite communication technology has a positive influence on television news coverage. However, that study was based on only five years of data, representing the period from 1972 through 1976, for one network, CBS. The present research, covering 10 years, and utilizing data from two networks, ABC and CBS, shows no positive relationship between Intelsat connections and television news coverage. The best explanation for this finding is that Intelsat has become virtually a global system, with earth stations in nearly all of the major non-Communist nations of the world. While satellite communication has taken over the role formerly served by cable, it does not influence the locations of news coverage by the U.S. networks.

Of the three independent variables investigated in this chapter, two have a statistically significant relationship with television news coverage. They are the location of network bureaus and the presence of AP or UPI in a nation. In a multivariate analysis, the relationships remains statistically significant when controlling for both the population of nations and their Gross national Product Per Capita.

Although two of the independent variables show a statistically significant relationship to television news coverage, network bureau location is a much stronger predictor. This finding is consistent with the worldwide and regional patterns of network television coverage described in earlier chapters.

Chapter 6

Television News and the Foreign Policy Process

More than a decade ago, Theodore White succinctly described the agenda-setting power of the press as follows: "No major act of the American Congress, no foreign adventure, no act of diplomacy, no great social reform, can succeed in the United States unless the press prepares the public mind" (White, 1973, p. 327). In the 1980s, television may well be the most influential part of the press in relation to the public mind or public opinion, particularly on international issues.

In this chapter the focus of attention shifts to the role and influence of network television news in relation to the foreign policy process. This focus on political aspects of television's international affairs coverage rather than such legitimate concerns as international education or audience comprehension of international news is deliberate. It stems from a conviction that the mass media, and in particular television, have become an important and powerful instrument in today's process of foreign policy formation and public diplomacy.

This view corresponds with much anecdotal evidence, accumulated during the 1970s and early 1980s. It also represents a growing conventional wisdom among some professional journalists, government officials, political scientists and communication researchers. Despite this rather widespread acknowledgment of television's power in the political process, social science research findings that convincingly demonstrate such power and influence remain as elusive as ever. Consequently, this chapter will blend some of the anecdotal observations with the available research findings in an effort to outline the key aspects of television's role in the foreign policy process. It purposely contains more conjecture than Chapters 2 through 5, which were all based on the analysis of empirical research data. Where possible, it identifies gaps in our current understanding of television news and foreign policy, suggesting possible directions for future research.

The organizational framework for this chapter is drawn from Cohen's widely cited and influential study *The Press and Foreign Policy* (1963). The question he addressed was the following.

"What are the consequences, for the foreign policy-making environment, of the way that the press defines and performs its job, and of the way that its output is assimilated by the participants in the process?" (Cohen, 1963, p. 4)

Although Cohen acknowledged a broad definition of the press, which included radio and television as well as the print media, as a practical matter he gathered most of his information concerning newspapers. One major reason was that, in the days before television news archives, the products of print media were less ephemeral and therefore easier to study at leisure than the output of the broadcast media.

A major contribution of Cohen's (1963) study was the clear identification and analysis of three major roles of the press in the foreign policy field. The first of these is the role of the press (here television news) as *observer*, which focuses on important aspects of the search for and presentation of foreign policy news. The second is the role of the press as *participant*, focusing on the interplay between the press, policymakers and other participants in the foreign policy process. Finally, the third role is that of the press as *catalyst*, looking at the manner in which the press is used by the public to satisfy its interests in foreign affairs and the implications of this role for foreign policy coverage. Cohen (1963) acknowledges that these three roles are not mutually exclusive, but they come close to being exhaustive. Taken together, they define what the press does in the foreign policy process. The following pages examine each of these three roles as they apply to network television coverage of international affairs during the 1970s and into the 1980s.

NETWORK TELEVISION AS AN OBSERVER
OF FOREIGN POLICY NEWS

Network television news today occupies a far more important role as an observer and conveyer of news about foreign affairs than it did at the start of the 1960's when Cohen (1963) authored his important study. As explained in earlier chapters, this evolution of television as a key medium in relation to the foreign policy process is tied to the rapid development of satellite communication and electronic newsgathering technology. Yet it is not technology alone that tells the story, but also the social control of such technology by the news media, along with governmental and nongovernmental policymakers.

The following sections will discuss three ways in which the changing technology of television newsgathering has influenced the foreign policy process. The first is the trend toward live or same-day coverage of war and conflict. The second is the response to television news reporting elicited from some governments. Finally, the third is the impact of the new technology on the nature and practices of television's foreign correspondents.

War and Crisis Reporting Via Satellite

The Vietnam War was the first prolonged conflict to receive sustained coverage on network television. During the late 1960s and early 1970s scenes of fighting in Indochina or related political maneuvering came into American households on a regular basis. However, coverage of the Vietnam war was primitive by comparison to coverage of later conflicts. Some of the visual coverage was shipped by plane to the United States and even if sent via satellite, most pictures were at least 24 hours old by the time they reached American television sets (Epstein, 1974; Mosettig and Griggs, 1980).

As mentioned in Chapter 1, the 1973 Arab–Israeli war in the Middle East marked a watershed in network television coverage of overseas conflicts (Fenton, 1980; Mosettig and Griggs, 1980). Israel possessed a satellite earth station and the equipment to develop the 16-millimeter color film that was still in use at that time. This allowed viewers in the United States to receive same-day coverage, direct from the war zone (Fenton, 1980).

During the 1970's and early 1980's a number of conflicts received live or same-day coverage by the U.S. networks. They included wars in Nicaragua, El Salvador and Lebanon, all of which directly involved United States political interests. However, during the same period of time there were conflicts that could not be covered on a live or timely basis because of political constraints. These included the Soviet invasion of Afghanistan and the bloody revolution in Cambodia between 1975 and 1979 (Adams and Joblove, 1982), and the war between Iran and Iraq. Logistical problems and British government policy prevented same-day coverage of the actual fighting in Britain's war with Argentina over the Falkland Islands.

From a purely technical standpoint, it will soon be possible to broadcast live coverage of wars and conflicts from virtually anywhere in the world. A producer for the NBC Nightly News has sketched a somewhat futuristic scenario in which live coverage of war is brought into American living rooms, with reporting from both sides of the battle (Wolzien, 1980). What would be the political and journalistic implications of such television news reporting? Evidence is already available that some, if not most governments take the political implications into consideration in their policies toward television news reporting.

Governmental Control and the New Technology

As indicated earlier, improved electronic newsgathering technology, satellites, and improved air travel connections do not automatically add up to more extensive visual coverage of news from different parts of the world. Governments can delay or prevent the transmission of a television news report in a variety of ways witout resorting to censorship as traditionally practiced in earlier years. For example, visas can be denied, cameras and editing equipment can be delayed at customs, or television correspondents can be denied use of a satellite earth station for feeding the story back to the United States.

Numerous specific instances of such tactics could be cited during the period of time covered by the present study. Some were mentioned briefly in Chapter 1. American correspondents were expelled from Iran immediately after the regime of the Ayatollah Khomeini took power, and later were readmitted after the seizure of hostages at the American embassy (Mosettig and Griggs, 1980). In 1980 during the war between Iran and Iraq, the Iraqi government refused television reporters access to Iraq's satellite facilites. As a result, videotape had to be driven 16 hours across the desert to Amman, Jordan for transmission to the United States. The delay meant that pictures seen on American television were at least one day old (Townley, 1981). According to Mosettig and Griggs (1980), the networks would cover Saudi Arabia more fully than they do now if the Saudis encouraged foreign press visitors or allowed the establishment of news bureaus. They also state that "Many black African nations will not allow the networks to visit even briefly. The problem of access to Third World countries is likely to grow worse, not better" (Mosettig and Griggs, 1980, p. 78).

A New Role for Television's Foreign Correspondents

The new technology of television newsgathering, with its capability of covering "hard" news such as war, political crises or natural disasters, has helped to create a new generation of foreign correspondents, with different training and a different approach to gathering international news for television. According to Mosettig and Griggs (1980), these new foreign correspondents are

> the "firemen" who jet from crisis to crisis. Sometimes they are based in the United States. The skills they bring to the job include the ability to work fast and turn out stories under pressure.

> This premium on agility, stamina, and speed worries some veteran foreign correspondents like John Chancellor of NBC. He feels that the technology has overcome correspondents; and while it helps them to cover more it hurts

them in terms of depth. Correspondents of his generation were expected to live in a country and know its language, culture and history. Now, he says, a group of mobile correspondents without that background is covering stories of increasing complexity. (Mosettig and Griggs, 1980, p. 71)

This assessment is shared by other professionals in television journalism. Veteran foreign corresondent Charles Collingwood has poignantly described a period during the 1940s and 1950s when foreign correspondents were expected to have a thorough grasp of the politics, history and peculiarities of the nation in which they were stationed. There was occasional travel involved, to cover summit meetings or international conferences, but most of the correspondent's time was spent cultivating knowledge of one nation and reporting from it. Collingwood's description of today's foreign correspondent is in stark contrast to his own earlier experience.

> Today's foreign correspondent—especially a television foreign correspondent —is a peripatetic figure, crisis-oriented, likely to be sent anywhere at any time, dispatched by the dictates of breaking news as interpreted by producers and editors in his home office. Of course, he has to be based somewhere, but his base is no longer his primary responsibility.

> His primary responsibility is to be ready to take off at a moment's notice to cover the latest crisis, which is usually in an unfamiliar place, with an unfamiliar cast of characters and whose troubles are ambiguous at best. (Collingwood, 1980, pp. 6, 8)

What are the consequences of sending television reporters overseas to report on a part of the world with which they have little experience? One of the most dramatic recent examples was the case of ABC correspondent Bill Stewart, who was sent from the network's New York bureau to cover the civil war in Nicaragua. Although Stewart had previously reported from Iran, colleagues noted that he was inappropriately trained for reporting from Latin America. He spoke no Spanish and had no background in the region. As a result, Stewart hired as an assistant a young Nicaraguan who was suspected of being a Sandinista. The youth was apprehended and shot at a government roadblock, and moments later correspondent Stewart was executed. The entire episode was captured on videotape by Stewart's cameraman, and all three U.S. television networks broadcast stories, including film of the shooting (Weisman, 1983). It is difficult to overestimate the political impact of that single incident, for the Somoza government in Nicaragua and its relations with the United States government. Some claim that the event accelerated the fall of the Somoza government and certainly made it difficult if not impossible for any U.S. administration to mobilize the public support that would be necessary for pro-Somoza policies. Westin (1982) noted that

On the anniversary of his [Stewart's] death, the government of Nicaragua staged a memorial service for him in the belief that his death played a large part in galvanizing world opinion to oppose the dictatorial regime of ousted President Somoza. (p. 120)

The background and training of television's foreign correspondents have changed somewhat since Cohen's (1963) study. For example, Batscha (1975) reports that 49 percent of network foreign correspondents hired before 1963 had newspaper experience compared to 38 percent who came from a television background. Among foreign correspondents recruited after 1963, 48 percent had a television background and only 38 percent newspaper training (Batscha, 1975, p. 4). Batscha (1975) also indicated that the trend toward hiring foreign corresondents with a television background was continuing. A new development that may accelerate that trend is the increased reporting from overseas by local television news correspondents from major U.S. markets. Local television stations from such markets at New York City, Los Angeles, Chicago and Minneapolis have begun to dispatch their correspondents overseas, particularly when there is some local connection with the international news event. Such connections are numerous in large, ethnically diverse cities of the U.S. (Townley, 1982).

Another possible difference between today's foreign correspondent for television and those of earlier years involves the level of interest in a particular nation or region of the world. In a study of U.S. journalists reporting from Latin America, Pollock (1981) reported that 6 out of 10 said they liked reporting on Latin American not because of its inherent interest as a region but because of the challenges it offers reporters as professional news gatherers. Batscha's (1975) investigation of why foreign correspondents for television entered the profession suggests that this finding is generalizable to most of network television's current foreign correspondents. In short, today's television journalists are generally more mobile and less interested in a single country or region of the world than were their predecessors.

NETWORK TELEVISION AS A PARTICIPANT IN THE FOREIGN POLICY PROCESS

The second role of television news is that of *participant* in the foreign policy process. Cohen (1963, p. 133) rejected the mirror metaphor which suggests that the press holds up a mirror in which its readers can see the world. Instead, he suggested that foreign affairs news comes out of an interplay between policymakers and international affairs correspondents.

At the time of Cohen's (1963) study, this interplay took place primarily through certain elite newspapers which placed more emphasis on foreign affairs reporting than did other media. He analyzed policymakers' views of the press, their contributions to international affairs coverage in the press, and their characteristic uses of such coverage. Although the elite newspapers and the major wire services remain important, in the two decades since Cohen's (1963) study television has assumed a new and more important role in the foreign policy process. It has taken its place alongside the major print media as a participant in the foreign policy process. The following sections look at the interplay between television news and policymakers in two ways. First, they examine the use of television news output by foreign policy elites and next look at the same policymakers' inputs into such coverage.

Use of Television News by Foreign Policy Elites

Despite the availability of official channels of information including reports by government agencies and intelligence networks, most foreign policy elites depend heavily on normal news channels for the bulk of their information about events around the world. Even intelligence reports themselves and diplomatic dispatches are heavily dependent on the major news media (Davison, 1974, pp. 16, 17). Top foreign policymakers from the president on down are generally heavy consumers of the press. Cohen (1963) identified the elite press, including the *New York Times* and *The Washington Post*, along with AP and UPI as principal sources of news for officials at the White House, State Department, Defense Department and Capitol Hill. In addition, he noted that individual correspondents for these leading print media would often be personal sources of information through their contacts with government officials.

While the elite press and major news agencies remain important sources of day-to-day information for government officials, network television news must surely be added to the list. President Lyndon Johnson had three television monitors installed side-by-side in the oval office of the White House so that he could monitor all three television networks simultaneously. Apparently President Nixon relied more heavily on a daily written summary of press output. Because of its political impact, all presidents now monitor network television news in some way or another, as do policy elites in other branches of government.

The relative importance of television news in comparison to the print media relates in part to the urgency or importance to policy elites of immediate or timely news reports. The inauguration of President Reagan on January 21, 1981, the same day that American hostages were returned from Iran, provides an excellent case in point. All three of the U.S. television networks were wired and equipped to report not only the inauguration activities in Washington, D.C., but also the return of U.S. hostages from Iran via Algeria. Both reporters and officials present at the inauguration were able to monitor events that transpired half a world away by means of small television monitors used by the network news teams. For some of them, television was able to convey what was happening more quickly than official channels (Diamond, 1982).

Paletz and Entman (1981) suggest that the mass media, including television, foster a dialogue among policy elites that in turn influences the bulk of the American public. Despite the preceding evidence that television news content plays an important role in the foreign policy process, there remains a dearth of research into its changing role.

Contributions to Television News by Foreign Policy Elites

The policymakers who most directly affect the foreign policy of the United States not only utilize the content of network television news, but also influence and contribute to that content in several ways. News conferences or briefings at the White House, State Department, Pentagon, or on Capitol Hill are usually certain to receive some network television coverage. Overseas travel by the president, secretary of state, secretary of defense or other senior officials is certain to generate network television coverage.

The general tendency of news media to focus attention on elites has been hypothesized and to a great extent documented (Galtung and Ruge, 1965). Gans (1979) estimated that the principal actor in about 75 to 80 percent of domestic stories on U.S. television news and in U.S. news magazines was a well-known or important individual. In the realm of international affairs, the president of the United States is the single most important newsmaker in the eyes of major media, followed by the secretary of state and certain other high government officials or members of Congress. This explains in large part why covering the White House is considered the most prestigious reporting assignment in network television.

The president has at his disposal a number of means to generate news coverage by television and other media. One is overseas travel. Wherever the president goes, representatives of the major news media are sure to follow. As an alternative, the president may dispatch the secretary of state, secretary of defense, or a special ambassador to some nation or region of the world and again, network television reporters will follow.

Another important means by which the president can assure news coverage is the presidential news conference. News conferences can be held at the White House or on the road when the president is traveling. In a recent study, Storey (1983) examined the presidential news conference as a vehicle for transmitting international affairs information to the public through network television. His research spanned the presidencies of

Nixon, Ford, Carter and Reagan. Using published transcripts of sampled presidential news conferences and the *Television News Index and Abstracts*, he compared international issues as treated in the press conferences and on network television's early evening news broadcasts.

In general, Storey (1983) found that there was a positive relationship between presidential emphasis on international issues in news conferences and their coverage on network television news. Although geographical focus of presidential discussion shifted from Vietnam and Indochina during the Nixon and Ford administrations to Egypt, Israel and the Middle East for President Carter, and toward more emphasis on the Soviet Union with President Reagan, the general tenor of presidential comments showed consistency across the four presidents.

Approximately one-third of international issues raised in presidential news conferences received coverage on the major network news broadcasts. The majority of these were reported directly by White House correspondents during the early portion of the network news broadcast, an indication of their importance relative to other news items. More than three-quarters of the people mentioned in television news stories were heads of state or other high-ranking government officials. In terms of news geography, Storey (1983) found no major or dramatic differences between presidential news conference comments and network television coverage of those remarks. The focus of both was oriented largely toward the Middle East (including North Africa), Asia, and Eastern Europe (including the USSR).

The general finding of consonance between presidential comments on international affairs issues and network television reporting of those issues is significant for what it implies about the participatory role of the press (here network television) in the foreign policy process. It lends empirical support to the notion that the president and the media tend to view foreign affairs from a common perspective and are more often in a collaborative than an adversary role.

An early and powerful example of the media impact of presidential travel was President Nixon's visit to the People's Republic of China in 1972. According to Henry Kissinger, President Nixon's former national security adviser and secretary of state, President Nixon was "Very conscious of the impact of television. He always thought that print journalism was not very important, but that television journalism was" (Hickey, 1983, p. 2). As a result, television reporters won a disproportionate share of space on the White House press list for Nixon's China visit. The arrival in Peking was carefully timed to occur at 11:30 on Monday morning so that it would appear live, via satellite, on American television Sunday night (Hickey, 1983, p. 2).

ABC's weeknight news coverage before and during Nixon's visit to the People's Republic of China in the fall of 1972 illustrated the importance attached to television news. As at the other networks, coverage of plans for the visit began many days before the president departed Washington for China. The sequence of preliminary reports is illuminating.

About a week or more before the president's departure, Harry Reasoner gave this introduction to a story on the forthcoming Nixon visit.

> Three planes, carrying some 100 communications technicians and White House officials arrived in Peking today to prepare for President Nixon's visit, which begins the 21st of this month. Today in Washington, Mr. Nixon referred to his coming trips to Peking and Moscow, as he spoke to government and civic leaders at the annual National Prayer Breakfast.

Following that introduction, ABC showed a videotape containing portions of President Nixon's speech in which he discussed the purpose of his trip to China.

The next weeknight news report on ABC featured correspondent Tom Jarriel, reporting from the White House on arrangements for the China trip. His report was based on additional details about the trip which had just been released by the White House. He discussed arrangements for Mr. Nixon's air and ground travel while in China, and the reduced size of the entourage which would accompany the president, noting that the press corps would be only one-third its normal size. The remainder of the report was a description of the main members of the official party traveling with the president. When he came to the presidential press secretary correspondent Jarriel had this to say:

> The administration's front man is Ronald Zeigler, the Presidential Press Secretary, who reflects his boss like a carbon copy. The importance attached by the White House to news coverage of this trip is indicated by Zeigler's fourth place ranking in the official party of 13.

Next, ABC sent its science editor, Jules Bergman, to do a special report on the president's specially equipped Boeing 707 named Air Force One. Bergman reported from inside Air Force One, describing in detail the layout of the plane and how its communication capabilities would be used on the China trip. Eight newsmen from the White House pool, he noted, would be among the passengers on Air Force One.

The day before the president's departure for China, ABC reported that he spent most of the day at Camp David, reading about China in preparation for the trip. The network also described bipartisan praise for the president's trip from members of Congress.

On the day of the departure, ABC carried live reports on the departure ceremonies at both the White House and Andrews Air Force Base. Correspondent Sam Donaldson, concluding the reports from Washington, noted that three planes would be accompanying Air Force One to China and that there would be a press contingent of 87 reporters in all.

On the president's first day in China, Harry Reasoner reported from Peking on the day's events. His report included a still photo of a surprise one-hour meeting between President Nixon and Chairman Mao. The most extensive videotape in the report showed a dinner in the Great Hall of the People during which Premier Chou En Lai and President Nixon exchanged toasts. Chou's toast included, in part, the following:

> The social systems of the U.S. and China are fundamentally different, and there are great differences between the Chinese government and the United States government. However, these differences should not hinder China and the United States from establishing normal state relations. We hope that through a frank exchange of views between our two sides to gain a clearer notion of our differences and make efforts to find common ground. A new start can be made in the relations between our two countries. In conclusion, I propose a toast, to the health of President Nixon, and Mrs. Nixon, to the health of our other American guests, to the health of all our friends and comrades present, and to the friendship between the Chinese and the American people.

Reasoner noted, following videotape of the toast, that no one had expected such candor, warmth, and commitment from a leader whose party may not be as enthusiastic about friendship with America. The report continued with the following toast by President Nixon:

> You said in your toast, the Chinese people are a great people. The American people are a great people. We have at times in the past, been enemies. We have great differences today. What brings us together, is that we have common interests, which transcend those differences. And so let us, in these next five days, start a long march together, not in lock step, but on different roads leading to the same goal, the goal of building a world structure of peace and justice in which all may stand together with equal dignity, and in which each nation, large or small, has a right to determine its own form of government, free of outside interference or domination. I ask all of you present to join me in raising your glasses to Chairman Mao, to Prime Minister Chou, and to the friendship of the Chinese and American people, which can lead to friendship and peace for all people in the world.

The ABC broadcast concerning the first day's events in China also included a report on Nixon's meeting with Chairman Mao. The network television visual reports via satellite continued on a daily basis throughout the remainder of the Nixon visit. The visit was both an historic political event and a watershed of sorts for the U.S. television networks in their coverage of presidential travel.

TELEVISION NEWS AS A CATALYST IN FOREIGN POLICY

The third and final role of television news in relation to foreign policy is that of *catalyst*. Cohen (1963) examined the way in which the press was used by the public to satisfy its interests in foreign affairs and the implications of that role for foreign policy coverage. As Cohen (1963) and others (Paletz and Entman, 1981) observe, it is a small number of "attentive" members of the public, including those elites who have a direct influence on policy, who are the key to understanding television's role as a catalyst in relation to foreign policy. Particularly in recent years, one of the major conceptual approaches used in studying the effects of mass media is the agenda-setting approach. The concept appears to have special relevance to the assessment of the influence of television news in relation to foreign policy.

The Agenda-setting Hypothesis

Cohen (1963) is usually credited with being one of the first to state the agenda-setting hypothesis which has stimulated so much research by communication scholars and political scientists. He characterized the press as follows: "It may not be successful much of the time in telling people what to think, but it is stunningly successful in telling its readers what to think *about*" (Cohen, 1963, p. 13). Today network television is arguably the most powerful agenda-setting medium in the United States.

The concept of agenda-setting has been operationally defined by McCombs as follows:

> The idea of agenda-setting influence by the mass media is a relational concept specifying a postive-indeed, causal-relationship between the emphases of mass communication and what members of the audience come to regard as important. In other words, the salience of an issue or other topic in the mass media influences its salience among the audience. (McCombs, 1981, p. 126)

Although the agenda-setting hypothesis is far from universally accepted as a conceptual approach to the study of mass media (Gandy, 1982) it has generated a great deal of research. Furthermore, it is particularly attractive as a conceptual approach to television news about international affairs, for several reasons.

First, one approach to research on agenda-setting has concentrated on making explicit the contingent conditions which might either enhance or restrain the agenda-setting role of the mass media. Such contingencies may arise both from the nature of the issues on the agenda of press and public, and from the characteristics of mass media audiences. One such contingent condition is the obtrusiveness of an issue, with obtrusive issues defined as those which are directly and personally experienced by individuals, while unobtrusive issues are almost exclusively in the domain of mass media (McCombs, 1981, p. 132). Using this approach it is clear that the bulk of international affairs news for most Americans would be categorized as unobtrusive issues. It has been well established that the average American obtains most knowledge of international affairs from the mass media, since they are for most people quite distant from personal experience (Davison, 1974, p. 9). Even among foreign policy elites one would expect to find widely varying degrees of familiarity with events in different parts of the world. As Cohen (1963) has noted, many personnel in the State Department are area specialists, who concentrate on and communicate with one area of the world.

Second, for the American public at large, a case was made in Chapter 1 that television is the dominant source of international news. Unlike some of the policy elites, the mass public does not read the *New York Times* or *Washington Post,* nor does it have access to the AP and UPI wire services on a continuing basis. Because of the scant attention to foreign news by most local newspapers, along with local television stations which mostly rebroadcast network reports of international news, the three television network news organizations are probably the dominant agenda-setting influence on the general public, at least when it comes to international issues.

A third reason that the agenda-setting framework appears especially appropriate for the study of television's influence in the foreign policy area is the present lack of empirical research in this domain. The following section comments on some of the existing agenda-setting research and the potential for future research that more directly addresses the question of television's influence on the foreign policy process.

Empirical Research on Agenda-setting

From a historical perspective, it is helpful to remember that communication research, as a behavioral discipline, was born in the years during and immediately following World War II, a period of intense concern with the effects of Nazi propaganda. The first two decades or more of communication research produced findings which showed that the mass media did not have powerful direct effects on beliefs and attitudes. Instead, they had limited or minimal influence, generally reinforcing rather than changing the public's opinions and attitudes.

The voluminous research on agenda-setting in the 1970's was in large measure a response by behavioral scientists who were disappointed with the minimal or limited effects model that had been so widely accepted. Much of the empirical research on agenda-setting to date has concentrated on political communication during election campaigns. McCombs (1981) advocates an expansion of such research to other substantive domains, including noncampaign settings. The possible agenda-setting influence of the major news media, including network television, in the domain of foreign policy issues and priorities would appear to be ripe for further research.

Another difficulty with the existing research on agenda-setting is that much of the research is cross-sectional in nature. By contrast, the hypothesized agenda-setting function of the media is dynamic in nature. Issues appear and disappear over time and hypothetically, as the media give an issue more attention or greater salience, the public gives the same issue more attention or rates it as more important. A recent experimental study of the consequences of television news programs (Iyengar, Peters and Kinder, 1982) suggested that the problem with much of the agenda-setting research may rest with the cross-sectional nature of the evidence, not with the hypothesis. The researchers suggested that, in order to be detected, agenda-setting effects must be investigated over time. They conducted two experiments in an effort to deal with this methodological problem. The following paragraphs provide a brief summary of their research.

In each of the two experiments, residents of the New Haven, Connecticut area were invited to Yale University to view videotaped network news broadcasts, ostensibly so that researchers could better understand how the public evaluates news programs. Participants were asked to watch the broadcasts at Yale "to ensure that everyone watched the same newscast under uniform conditions" (Iyengar, Peters and Kinder, 1982, p. 851).

Each of the experiments spanned six consecutive days. On the first day, subjects in the first experiment completed a questionnaire dealing with a wide range of political topics. After being randomly assigned to an experimental and a control group, the two groups of participants spent the next four days viewing videotaped news broadcasts. Unknown to them, the videotapes shown to the experimental group had been altered to provide sustained coverage of a particular national issue. The tapes viewed by the control group contained no coverage of that issue. The videotaped broadcasts were altered by using sophisticated editing equipment to insert stories provided by the Vanderbilt Television News Archive. On the final day of the experiment, the participants completed a second questionnaire that again included measures of the importance of political issues.

The issue chosen for sustained coverage in the first experiment was alleged weaknesses in U.S. defense capability. In the second experiment, three issues were used and participants were divided into three experimental groups. The issues were national defense preparedness, pollution of the environment, and inflation.

The results of both these experiments provided evidence supporting the agenda-setting hypothesis and suggesting that television viewers who are less able or willing to counter-argue with a news presentation may be more vulnerable to agenda-setting influence. These findings would suggest that the politically active public and certainly the foreign policy elites are less susceptible to any agenda-setting influence of television news than the public at large.

Political Implication of Agenda-setting for Foreign Policy

The agenda-setting conceptual framework implies that television news attention to foreign affairs has enormous political consequences, even though those consequences may be entirely unintentional on the part of media professionals. Policymakers are free to ignore or postpone consideration of items that are not on the public agenda (Iyengar, Peters, and Kinder, 1982; White, 1973).

In the context of the present study, particularly findings presented in Chapters 3 and 4 concerning coverage of developing nations, the implications of agenda-setting by network television are disturbing. Conspicuously absent from the world view provided by network television's window are the important phenomena of the ongoing struggle for social change and development in nations of Africa, Asia, and Latin America. Such "development news" seldom falls within the prevailing news selection criteria used by the American television networks as they gather and disseminate information about events around the world. In political terms, this lack of network news attention to issues of social change and development in the Third World gives U.S. policymakers broad leeway to ignore, minimize, or postpone consideration of such problems. Such a view implies that fundamental structural changes in the international news system, including a redefinition of what constitutes good television news coverage of world events, will be necessary if progress is to be made on the political front.

Chapter 7

What Sort of Window on the World?

Until recently the topic of international affairs coverage by the U.S. television networks has more often been the subject of debate and polemics than of empirical analysis. Frank (1973) devoted some attention to foreign affairs coverage and Batscha (1975) approached the topic from the viewpoint of reporters working for the television networks. However, except for some scattered articles in scholarly journals, the work of Adams (1981, 1982) represents the first effort to bring together studies that systematically analyze the international affairs content of network television. In part, this reflects the practical problems associated with videotaping television news broadcasts for content analysis, which were resolved with the establishment of the Vanderbilt Television News Archive in 1968.

What conclusions can be drawn or what lessons learned from the preceding analysis of over 7,000 international news items contained in more than 1,000 early evening network news broadcasts during the 1972–1981 decade? Hopefully one lesson implicit throughout the study concerns the value of systematic, empirical inquiry concerning television news, the sort of inquiry that working journalists, with the demands of their work, might find difficult to undertake. Although few, if any, of the findings reported in this study should come as a surprise to reporters, producers, writers and editors in television news, they may offer a chance for some reflection on the craft of journalism and some of its basic patterns as they relate to international affairs. As indicated in Chapter 1, the findings should also be of interest to policymakers and other researchers, both in the United States and abroad.

SUMMARY OF MAJOR FINDINGS

What were the characteristic features of network television's international news coverage during the decade covered by this study? One was that the extent and dimensions of international news coverage vary little from net-work to network. In terms of the major findings discussed below, there are virtually no cross-network differences large enough to be of interest. The broad, overall patterns of news coverage are remarkably similar across networks. The other main characteristics of international news on network television may be broadly summarized under four headings:

1. the amount of coverage,
2. the nature of such coverage,
3. influences on international affairs reporting, and
4. impact of television's international news coverage.

The Amount of Coverage

The findings of this study show clearly that television news devotes a sub-stantial proportion of its coverage to international affairs. This is true whether measured in terms of international news stories or the proportion of available broadcast time allocated for international news. On the aver-age, 7 out of 17 news items, or nearly 40 percent of those broadcast on a typical weeknight newscast, deal with international affairs. Of the 22 or 23 minutes available for news, about 10 minutes or 45 percent of avail-able air time will characteristically be devoted to international affairs. By either measure, this is a far higher proportion of international news than that which is printed in most newspapers. Only newsmagazines such as *Time* or *Newsweek*, or elite newspapers such as *The New York Times*, would carry proportionately as much international news.

Although the television networks do devote a significant proportion of their stories and air time to international affairs, with their early evening news broadcasts, they face the difficult problem of covering world events in 10 minutes. In addition, the intense competitive pressure among the networks and their concern with share of audience influence the produc-tion techniques used to assemble a half-hour news broadcast. Given such constraints and influences, what is the nature of the international affairs news carried by the networks?

The Nature of the International News Coverage

Chapters 2 through 4 described some prominent dimensions of network television's international news coverage, including the format of presenta-tion, crisis content, news geography, and the nature of coverage given to developed, developing and socialist nations. The following are some of the key findings.

Presentation Formats

There was a definite trend toward greater use of domestic and foreign video formats and toward less use of anchor reports during the 1972–1981 decade. While this shift does indicate more direct reporting of international affairs by the networks and a lower level of reliance on news agency output, it is also a trend which occurred during a period of changing newsgathering technology, increased profitability for network news, and growing competition. Such factors undoubtedly contributed greatly to the changes in reporting format.

Crisis Content

Approximately 27 percent of international news broadcast by the networks dealt with crisis themes such as civil unrest, war, terrorism, coups, assassinations or disasters. However, the proportion of such crisis themes was highest among foreign video reports and lowest in domestic video reports, providing some support for the notion that television news has a strong appetite for dramatic visual material. For crisis news from areas of the world not immediately accessible to their correspondents, the networks tend to rely on news agency material, which is broadcast in the form of anchor reports.

News Geography

The major feature of network television's news geography is a pattern of hierarchy and extreme concentration of news coverage in which a small number of nations account for a very large proportion of all coverage. This pattern holds true both worldwide and within major geographical regions. Network news attention is concentrated on a small proportion of nations in the world and on a relatively small number of nations in each major geographical region. The concentration of coverage is most pronounced in the network's use of foreign video reports. Two regions, Western Europe and the Middle East, alone account for about two-thirds of all foreign video reports aired by the networks. Adding Asia explains the origin of fully 80 percent of such reports. The remaining foreign video coverage came from Eastern Europe and the USSR, Latin America, Canada, and Africa.

Roughly the same pattern holds true in terms of total references to each region in the news. Africa receives less attention than other regions, with Latin America not far ahead.

Along with the regional differences in amount of coverage, there are differences in the relative use of the three major reporting formats. The networks rely more heavily on the news agencies for coverage of Africa than for any other region, and they depend least on the agencies for Middle East news. Among other things, such a pattern reflects the concentration of network bureaus in the Middle East and lack of them in sub-Saharan Africa. Latin America and Asia are reported relatively more often than other regions in domestic video reports, corresponding with U.S. involvement in Vietnam early in the decade and other direct foreign policy concerns in Latin America over the 10 years. Foreign video reports, as already summarized above, showed a definite regional pattern, with Western Europe, the Middle East, and (because of Vietnam) Asia, dominating the picture.

Regional differences also appear in crisis reporting, with two regions showing notable departures from the global pattern of crisis reporting. Africa was depicted as by far the most crisis-prone region of the world on network television news. Eastern Europe and the USSR, on the other hand, were portrayed with the smallest proportion of crisis items. In the case of Africa, South Africa is the only nation south of the Sahara that generates network television coverage with any consistency in the absence of a war or major crisis. The low level of crisis reporting from Eastern Europe and the USSR is due mainly to the heavy coverage of the USSR in the context of such topics as arms control, and global or regional diplomacy involving the superpowers. Government controls over newsgathering from that region of the world are another contributing factor.

All of the regional differences in network television reporting of international news illustrate the priorities of the U.S. networks during the decade from 1972 through 1981. However, they also appear to parallel major U.S. foreign policy interests over the same period of time.

Coverage of Developed, Developing and Socialist Nations

A comparative analysis of the coverage given to developed, developing, and socialist nations produced the following findings:

- In relation to total population and number of nations, developing countries receive less coverage than developed nations. They also receive less coverage than socialist nations, primarily because of heavy coverage given the USSR.
- The networks carry more foreign video reports from developed nations than from developing nations, and have increased foreign video coverage from socialist nations, while showing no increase in such coverage from developing nations.
- Developing nations are covered most often in connection with the U.S. or another developed nation and, in a related finding, developed or socialist nations are more likely than developing nations to be the exclusive focus of a news story.
- There is proportionately more crisis reporting from developing nations than from developed or socialist nations.

- The number of foreign video reports originating from developing nations did not increase over the 1972–1981 decade as it did for developed and socialist nations.

Some Influences on International Coverage

Chapter 5 looked at the relationship between elements of the international news communication infrastructure for television and network coverage of international affairs. Satellite communication channels, network bureau locations and the location of AP and UPI correspondents all correlate with network television's coverage of world news during 1972–1981. Network bureaus were the strongest influence shaping the coverage, even though news agency locations had some effect. Intelsat earth stations, although they are necessary for instant or timely reporting, were not an influence on network television's coverage.

The Impact of Television's International News

Chapter 6 did not produce clear-cut findings based on empirical data, in the same sense that earlier chapters did. However, the effects of network television coverage of international affairs, particularly in relation to the foreign policy process, are an essential part of the rationale for this and related studies. It would be difficult to imagine political scientists, sociologists, and communication researchers devoting so much time to the study of international news if such news had no important consequences. Even more difficult to imagine, if television coverage of international news had little impact, would be the behavior of heads of state, diplomats, and other government officials from many nations. Chapter 6 outlined some of the important ways in which the role of television news in relation to the foreign policy process changed during the 1970s.

The influence of network television in relation to the foreign policy process changed dramatically over the past 10 to 15 years. The change is evident in relation to each of the three roles of television news in the field of foreign policy: as observer, participant and catalyst.

The development of new technology for newsgathering spurred changes in the role of network television as an observer of international affairs. With lightweight cameras and videotape editing equipment, foreign correspondents became more mobile. At least partly due to such technology, a new breed of correspondents took over at the three networks, ready to report at a moment's notice from any location in the world with good airline connections and an accessible satellite earth station. The accessibility of satellite earth stations sometimes depended upon government policies. The list of nations that denied the networks permission for satellite transmission grew longer over the past decade.

Over the same time period, network television showed its increased importance as a participant in the foreign policy process. More than ever before, heads of state and other government officials used television as a channel for diplomatic messages. Sometimes, network reporters played an active role in the process, as when Walter Cronkite interviewed Egyptian President Sadat and Israeli Prime Minister Begin with the first public statement by both that Sadat would accept an invitation to visit Israel. In other instances, some critics claimed the networks played a less constructive role, as when one network broadcast an interview with an American hostage in Iran, accepting certain Iranian restrictions on the broadcast.

Although other nations use network television as a channel for public diplomacy, it is a principal forum for the president of the United States, the secretary of state and other government officials. The strong symbiotic relationship between television reporters and government officials showed no signs of weakening during the 1970s.

Finally, television news continued its role as a catalyst in the foreign policy process, understood here as its ability to place certain issues or concerns on the public agenda, or at least on the agenda of those elites who have more direct input into the policy process. To use the example of arms control, if questions about U.S. defense technology and expenditures are not on the public agenda or do not register as public concerns in opinion polling, policy elites have great leeway to continue with the *status quo*. On the other hand, if the major media do foster public concern, they may be an important catalyst in changing policies. As a national medium capable of attracting massive audiences so that diplomats from the United States and other nations can take their case directly to the public, television serves a unique and potentially powerful role in the foreign policy process. The shift of U.S. policy and public opinion toward Egypt during the presidency of Anwar Sadat is an intriguing example of the importance of television (Adams, 1982).

THE IMPLICATIONS OF THE FINDINGS

While the findings of this study do not provide easy prescriptions for the future, neither can they be dismissed without some comment on their importance or significance. Several aspects of the research deserve mention.

First, the research is a response to the need for a longitudinal approach in the study of international news content (Frank, 1973). Readers, including other researchers, can be confident that the major dimensions of television news content reported here represent stable patterns, and not transitory aspects of television news coverage. As such, the research offers a solid baseline measure of the amount and nature of international news coverage on U.S. network television. Such a baseline is necessary for meaningful comparisons with other media, in the U.S. or other nations, and for assessing the future performance of the television networks in this important area.

Second, the study reveals no clear trend, over the 10-year period 1972–1981, toward more broadly-based coverage, involving more nations, especially developing ones. With the present structure of the network news organizations, there may be a built-in limit to what they can accomplish, given commercial imperatives and a half-hour time slot for their major weekday newscasts. At the same time, it is worth noting that some viewers may be dissatisfied with the brevity of much international news reporting, which may seem accentuated with the interruption and pace of the numerous commercials in each newscast. Individuals from nations that do not allow commercials during news broadcasts may appreciate this comment more fully than some U.S. viewers.

Perhaps this finding that the networks continue to be oriented toward hard news, crises, and the visually exciting or audience-holding format is the least surprising of them all. However, it is well to remember that alternatives, such as expanded air time and news pooling to avoid duplication of effort across the networks, have been publicly discussed for years. Such practices would, almost by definition, be part of an effort to provide broader coverage of world events.

A third implication of this study is one step removed from the data that have been collected and the study's empirical findings. Given the present international concerns about news, particularly those voiced by Third World nations, network television may be able to solve some of its problems only through a broader definition of what constitutes news, and some basic structural changes. "ABC News Nightline," and some of the other expanded news programming of the networks may represent a move in that direction. However, the eventual question is whether the "development news" called for by Third World leaders can be accommodated by Western media, including television. Can television portray the social and economic processes that are occurring in other parts of the world and often at a pace that may seem agonizingly slow? Can it convey the urgency of the struggle for development in areas like health, education, and agriculture? Perhaps most important is the question of whether television can communicate to the American public how such problems involve them. What is the nature of the dependence or interdependence of nations and why is it important to the United States? Many would hope that television is capable of playing an important role in such communication.

Finally, it is significant and more than a little ironic that a broadened conception of televised international news is in the long-term self-interest not only of Third World nations, but also of developed nations like the United States. In an interdependent world it would seem to be of utmost importance that the United States seek to foster a broader and deeper understanding of the struggle for social change and economic development underway in most nations of the world. Even from a perspective of pure self-interest, the U.S. has a stragetic interest in many nations of the developing world. Many are sources of vital natural resources, and the dependence of the U.S. economy on international trade continues to grow.

Television is a national medium, and in the United States it is the principal source of international news for most of the public as well as a forum for debate on major international issues. Given that role, and despite the peculiarities of this particular medium, the question for the future is whether television's window can be changed to provide a more open and inclusive view of the world, and hence a better understanding of the problems and prospects facing all of its nations.

References

Adams, William C., (1978). "Network News Research in Perspective: A Bibliographic Essay." In William Adams and Fay Schreibman, eds., *Television Network News: Issues in Content Research*. Washington, D.C.: School of Public and International Affairs, George Washington University, 11–46.

Adams, William C., ed. (1981). *Television Coverage of the Middle East*. Norwood, New Jersey: Ablex.

Adams, William C., ed. (1982). *Television Coverage of International Affairs*. Norwood, New Jersey: Ablex.

Adams, William C. and Michael Joblove. (1982). "The Unnewsworthy Holocaust: TV News and Terror in Cambodia." In William C. Adams, ed., *Television Coverage of International Affairs*. Norwood, New Jersey: Ablex, 217–226.

Adams, William and Fay Schreibman, eds. (1978). *Television Network News: Issues in Content Research*. Washington, D.C.: School of Public and International Affairs, George Washington University.

Aggarwala, Narinder K. (1981). "A Third World View." In Jim Richstad and Michael H. Anderson, eds., *Crisis in International News: Policies and Prospects*. New York: Columbia University Press, xv–xxi.

Almaney, Adnan. (1970). "International and Foreign Affairs on Network Television News." *Journal of Broadcasting* 14, 499–509.

Altheide, David L. (1976). *Creating Reality: How TV News Distorts Events*. Beverly Hills, California: Sage.

Altheide, David. (1981). "Iran vs. U.S. TV News: The Hostage Story Out of Context." In William C. Adams, ed., *Television Coverage of the Middle East*. Norwood, New Jersey: Ablex, 128–157.

Batscha, Robert M. (1975). *Foreign Affairs News and the Broadcast Journalist*. New York: Praeger.

Bogart, Leo. (1975). "How the Challenge of Television News Affects the Prosperity of Daily Newspapers." *Journalism Quarterly*. 52 (Autumn), 403–410.

Bogart, Leo. (1980). "Television News as Entertainment." In Percy H. Tannenbaum, ed., *The Entertainment Functions of Television*. Hillsdale, New Jersey: Lawrence Erlbaum Associates, 209–249.

Bogart, Leo. (1981). *Press and Public: Who Reads What, When, Where and Why in American Newspapers*. Hillsdale, New Jersey: Lawrence Erlbaum Associates.

Boorstin, Daniel J. (1972). *The Image: A Guide to Pseudo-events in America.* New York: Atheneum.

Bower, Robert T. (1973). *Television and the Public.* New York: Holt, Rinehart and Winston.

Boyd-Barrett, Oliver. (1980). *The International News Agencies.* Beverly Hills, California: Sage.

Brody, Richard A. (1971). "Citizen Participation in Foreign Affairs: The Potential Effects of Public Opinion in American Foreign Policy." *Civis Mundi 4* (April).

Buergenthal, Thomas. (1974). "The Right to Receive Information Across National Boundaries." In *Control of the Direct Broadcast Satellite: Values in Conflict.* Palo Alto, California: Aspen Institute Program on Communications and Society, 73–84.

Chandler, Otis. (1977). Testimony before Subcommittee on International Operations of the U.S. Senate's Committee on Foreign Relations. *Congressional Record.* (June 15), S9804, S9805.

Chandler, Robert. (1982). Senior Vice-President, Administration, CBS News, Personal correspondence, (August 3).

Charles, Jeff; Larry Shore and Rusty Todd. (1979). "The *New York Times* Coverage of Equatorial and Lower Africa." *Journal of Communication.* 29(2), 148–155.

Cohen, Bernard C. (1963). *The Press and Foreign Policy.* Princeton, New Jersey: Princeton University Press.

Cohen, Bernard C. (1973). *The Public's Impact on Foreign Policy.* Boston, Massachusetts: Little, Brown.

Collingwood, Charles. (1980). "Prestige. Glamour. Access to High Places. Now that the role of TV's Foreign Correspondents has Changed, a Veteran bids... Goodby to All That." *TV Guide.* (April 19), 6–10.

Communications Satellites: Are the Users Ready? (1967). A Report on the Proceedings of a Conference at the Edward R. Murrow Center of Public Diplomacy, Tufts University, Medford, Massachusetts.

Comsat. (1980). *Comsat Seventeenth Annual Report to the President and the Congress.* Washington, D.C.: Communications Satellite Corporation.

Corrigan, William T. (1977). General Manager, News Operations, NBC News, personal correspondence, (July 12).

Crystal, Lester M. (1977). "Using Technology: Improve Substance, Not Just Appearance." *Television/Radio Age.* (October 24), 43–78.

Cutlip, Scott M. (1954). "Content and Flow of AP News—From Trunk to TTS to Reader." *Journalism Quarterly.* 31, 434–446.

Davison, W. P. (1974). *Mass Communication and Conflict Resolution: The Role of the Information Media in the Advancement of International Understanding.* New York: Praeger.

Davison, W. Phillips, Donald R. Shanor and Frederick T. C. Yu. (1980). *News From Abroad and the Foreign Policy Public.* HEADLINE Series 250. (August). New York: Foreign Policy Association.

Diamond, Edwin. (1982). *Sign Off: The Last Days of Television.* Cambridge, Massachusetts: The MIT Press.

Dizard, Wilson. (1966). *Television: A World View.* Syracuse, New York: Syracuse

University Press.

Edwardson, Mickie, Donald Grooms, and Susanne Proudlove. (1981). "Television News Information Gain from Interesting Video vs. Talking Heads." *Journal of Broadcasting.* 25(1), (Winter), 15–24.

Ehrlich, Ellen. (1977). Director of Information Services, CBS News, Personal Correspondence, (July 13).

Elliott, Philip and Peter Golding. (1974). "Mass Communication and Social Change: The Imagery of Development and the Development of Imagery." In Emanuel de Kadt and Gavin Williams, eds., *Sociology and Development.* London: Tavistock.

Epstein, Edward J. (1974). *News From Nowhere.* New York: Vintage Books.

The Europa Yearbook 1981: A World Survey. (1981). London: Europa Publications, Ltd.

Feders, Sid. (1978). Foreign Editor, CBS News, Telephone Interview in March.

Fenton, Tom. (1980). "Bringing You Today's War—Today." *TV Guide.* (November 15), 36–38.

Fishman, Mark. (1980). *Manufacturing the News.* Austin, Texas: University of Texas Press.

Frank, Robert S. (1973). *Message Dimensions of Television News.* Lexington, Massachusetts: D.C. Heath.

Friendly, Fred W. (1977). "Pooled Coverage: Small Step to TV News Breakthrough." In Ted C. Smythe and George A. Mastroianni, eds., *Issues in Broadcasting.* Palo Alto, California: Mayfield.

Friendly, Fred. (1982). "Let's Have an Hour of Network News." *Advertising Age.* (April 26), M-9.

Galtung, Johan. (1974). "A Rejoinder." *Journal of Peace Research.* 11(2), 157–160.

Galtung, Johan. (1979). "A Structural Theory of Imperialism." In George Modelski, ed., *Transnational Corporations and World Order.* San Francisco, California: W. H. Freeman and Company, 155–171.

Galtung, Johan and Mari Holmboe Ruge. (1965). "The Structure of Foreign News." *Journal of Peace Research.* 1, 64–91.

Gandy, Oscar H. Jr. (1982). *Beyond Agenda Setting: Information Subsidies and Public Policy.* Norwood, New Jersey: Ablex Publishing Corporation.

Gans, Herbert J. (1979). *Deciding What's News: A Study of CBS Evening News, NBC Nightly News, Newsweek, and Time.* New York: Pantheon Books.

Gerbner, George and G. Marvanyi. (1977). "The Many Worlds of the World's Press." *Journal of Communication* 27(1), 52–66.

Golding, Peter and Philip Elliott. (1979). *Making the News.* New York: Longman.

Harris, Phil. (1974). "Hierarchy and Concentration in International News Flow." *Politics* 9(2), (November), 159–165.

Harris, Phil. (1976). "Selective Images: An Analysis of the West African Wire Service of an International News Agency." Paper presented at the conference of the International Association for Mass Communication Research, University of Leicester, England.

Hester, Al. (1971). "An Analysis of News Flow from Developed and Developing Nations." *Gazette.* 17(1,2), 29–43.

Hickey, Neil. (1983). "Henry Kissinger and TV: 'Did I Sometimes Use the Press? Yes.'" *TV Guide.* 31(14), (April 2), 2–9.

Holsti, Ole R. (1969). *Content Analysis for the Social Sciences and Humanities.* Reading, Massachusetts: Addison-Wesley.

"How News Travels in a High-Speed World." (1977). *AP Log.* (January 31), 1,4.

Hulten, O. (1973). "The INTELSAT System: Some Notes on Television Utilization of Satellite Technology." *Gazette. 19*(1), 29–37.

Intermedia. (1981a). "TV News: Broadcasters Want a Better Deal." *Intermedia. 9*(4), (July), 82–83.

Intermedia. (1981b). "Asia's Special Needs for Satellite Links." *Intermedia. 9*(4), (July), 86–88.

International Commission for the Study of Communication Problems. (1980). *Many Voices, One World: Report by the International Commission for the Study of Communication Problems.* Paris: UNESCO.

Iyengar, Shanto; Mark D. Peters and Donald R. Kinder. (1982). "Experimental Demonstrations of the 'Not-So-Minimal' Consequences of Television News Programs." *The American Political Science Review. 76*(4), (December), 848–858.

Kastelnik, Connie. (1982). ABC News, personal correspondence, (July 12).

Katz, Elihu. (1973). "Television as a Horseless Carriage." In G. Gerbner and others, eds., *Communications Technology and Social Policy: Understanding the New "Cultural Revolution".* New York: Wiley, 381–392.

Katz, Elihu, Hanna Adoni and Pnina Parness. (1977). "Remembering the News: What the Picture Adds to Recall." *Journalism Quarterly. 54* (Summer), 231–239.

Katz, Elihu and George Wedell. (1977). *Broadcasting in the Third World: Promise and Performance.* Cambridge, Massachusetts: Harvard University Press.

King, Jonathan. (1981). "Visnews and UPITN: News Film Supermarkets in the Sky." In Jim Richstad and Michael H. Anderson, eds. *Crisis in International News: Policies and Prospects.* New York: Columbia University Press, 283–298.

Kliesch, Ralph E. (1975). "A Vanishing Species: The American Newsman Abroad." *1975 Membership Directory, Overseas Press Club of America.* 18–19, 110–125.

Lang, Kurt and Gladys Engel Lang. (1971). "The Unique Perspective of Television and Its Effects: A Pilot Study." In Wilbur Schramm and Donald F. Roberts eds. *The Process and Effects of Mass Communication.* Urbana, Illinois: University of Illinois Press, 169–188.

Larson, James F. (1978). "America's Window on the World: U.S. Network Television Coverage of International Affairs, 1972–1976." Unpublished Ph.D. dissertation, Stanford University.

Larson, James F. (1979). "International Affairs Coverage on U.S. Network Television." *Journal of Communication 29*(2), 136–147.

Larson, James and Andy Hardy. (1977). "International Affairs Coverage on Network Television News: A Study of News Flow." *Gazette 4*, 241–256.

Larson, James F. (1982). "International Affairs Coverage on U.S. Evening Network News, 1972–1979." In William C. Adams, ed., *Television Coverage of International Affairs.* Norwood, New Jersey: Ablex, 15–41.

Larson, James F. and J. Douglas Storey. (1983). *The East on Western Television: A Ten-Year Review of how the U.S. Networks Cover Asia.* Occasional Paper, Asian Mass Communication Research and Information Centre (AMIC), Singapore.

Lichty, Lawrence W. (1982). "Video Versus Print." *The Wilson Quarterly* 6(5), (Special Issue 1982), 49–57.

Lippmann, Walter. (1921). *Public Opinion.* New York: Macmillan.

Lippmann, Walter, and C. Merz. (1920). "A Test of the News." *The New Republic.* 23(296), 1–42.

McAnany, Emile G. (1975). "Television: Mass Communication and Elite Controls." *Society.* (September/October), 41–46.

McAnany, Emile G., James F. Larson and J. Douglas Storey. (1982). "News of Latin America on Network Television, 1972–1982: Too Little Too Late?" Paper presented at the annual conference of the International Communication Association, Boston, Massachusetts, (May).

McCombs, Maxwell E. (1981). "The Agenda-Setting Approach." In Dan D. Nimmo and Keith R. Sanders, eds., *Handbook of Political Communication.* Beverly Hills, California: Sage, 121–140.

Mankekar, D. R. (1981). "The Nonaligned News Pool." In Jim Richstad and Michael H. Anderson, eds., *Crisis in International News.* New York: Columbia University Press, 369–379.

Markoff, John and others. (1974). "Toward the Integration of Content Analysis and General Methodology." In David R. Heise, ed., *Sociological Methodology 1975.* San Francisco, California: Jossey-Bass, 1–58.

Matta, Fernando Reyes. (1976). "The Information Bedazzlement of Latin America." *Development Dialogue.* 2, 29–42.

Mills, C. Wright. (1959). *The Sociological Imagination.* London: Oxford University Press.

Mosettig, Michael and Henry Griggs, Jr. (1980). "TV At the Front." *Foreign Policy.* 38 (Spring), 67–79.

News Research Bulletin. (1974). "A Report to Client Newspapers of the Gallup Poll on the Credibility of the Press—Including Higher Circulation Prices." *News Research Bulletin 2* (February).

Nordenstreng, Kaarle. (1975). "Free Flow of Information: The Rise and Fall of a Doctrine." Paper presented at Department of Communication Colloquium, Stanford University, (April).

Nordenstreng, Kaarle and Tapio Varis. (1974). "Television Traffic: a one-way Street?" *UNESCO Reports and Papers on Mass Communication,* (70).

Ostgaard, E. (1965). "Factors Influencing the Flow of News." *Journal of Peace Research 1,* 39–63.

Paletz, David L. and Robert M. Entman. (1981). *Media Power Politics.* New York: The Free Press.

Paletz, David and Robert Pearson. (1978). "'The Way You Look Tonight': A Critique of Television News Criticism." In William Adams and Fay Schreibman, eds., *Television Network News: Issues in Content Research.* Washington, D.C.: School of Public and International Affairs, George Washington University, 65–85.

Pasadeos, Yorgo. (1982). "International News Coverage in U.S. Newsmagazines: A Content Analysis and Some Correlates." Ph.D. Dissertation, The University of Texas at Austin, Austin, Texas, (December).

Pearce, Alan. (1980). "How the Networks Have Turned News into Dollars." *TV Guide.* (August 23), 7–12.

Plante, James F. (1983). Director, News Services, NBC News. Personal Correspondence, (January 12).

Pollock, John C. (1981). *The Politics of Crisis Reporting: Learning to be a Foreign Correspondent.* New York: Praeger.

Powers, Ron. (1981). "Hour-Long Network News Is: a. Necessary b. Too Expensive c. Imminent d. All of the Above e. None of the Above." *TV Guide.* (August 22) 5–8.

Quint, Bert. (1980). "Dateline Tehran: There Was a Touch of Fear." *TV Guide.* (April 5) 6–12.

Richstad, Jim and Michael H. Anderson, eds. (1981). *Crisis in International News: Policies and Prospects.* New York: Columbia University Press.

Roper, Burns W. (1981). *Trends in Attitudes Toward Television and Other Media: A Twenty-Two Year Review.* Report for the Television Information Office, New York, (April).

Rosenblum, Mort. (1979). *Coups and Earthquakes: Reporting the World to America.* New York: Harper Colophon Books.

Rosenblum, Mort. (1977). "Western Wire Services and Third World News Agencies: The Practical Aspects." Paper presented at the Conference on the Third World and Press Freedom, New York, (May).

Rosengren, Karl E. (1970). "International News: Intra and Extra Media Data." *Acta Sociologica. 13*(2), 96–109.

Rosengren, Karl E. (1974). "International News: Methods, Data and Theory." *Journal of Peace Research. 11*(2), 145–160.

Rosengren, Karl E. (1977a). "Bias in News: Methods and Concepts." Paper presented to the Mass Communication Division, International Communication Association Convention, Berlin, May.

Rosengren, Karl E. (1977b). "Four Types of Tables." *Journal of Communication. 27*(1), 67–75.

Schiller, Herbert I. (1973). *The Mind Managers.* Boston, Massachusetts: Beacon Press.

Schlesinger, Philip. (1979). *Putting 'reality' together: BBC News.* Beverly Hills, California: Sage.

Schramm, Wilbur. (1964). *Mass Media and National Development.* Stanford, California: Stanford University Press.

Schramm, Wilbur. (1968). *Communication Satellites for Education, Science, and Culture.* UNESCO Reports and Papers on Mass Communication, (53).

Semmel, Andrew K. (1977). "The Elite Press, the Global System, and Foreign News Attention." *International Interactions. 3*(4), 317–328.

Sheehan, William. (1977). Testimony before Subcommittee on International Operations, U.S. Senate Committee on Foreign Relations. *Congressional Record 123,* (June 15), S9805, S9806.

Siebert, Fred S., Theodore Peterson, and Wilbur Schramm. (1956). *Four Theories of the Press.* Urbana, Illinois: University of Illinois Press.

Signitzer, Benno. (1976). *Regulation of Direct Broadcasting from Satellites: The U.N. Involvement.* New York: Praeger.

Somavia, Juan. (1976). "The Transnational Power Structure and International Information." Paper presented at a seminar on the role of information in the new international order, Latin American Institute for Transnational Studies, Mexico

City, (May).

Statistical Yearbook 1978. (1979). Thirtieth Issue, United Nations. New York: Publishing Service United Nations, 430–431.

Stauffer, John, Richard Frost and William Rybolt. (1981). "Recall and Learning from Broadcast News: Is Print Better?" *Journal of Broadcasting. 25*(3), (Summer), 253–262.

Stauffer, John, Richard Frost and William Rybolt. (1983). "The Attention Factor in Recalling Network Television News." *Journal of Communication. 33*(1), (Winter), 29–37.

Stevenson, Robert L. and Richard R. Cole. (1980). "Patterns of World Coverage by the Major Western Agencies." Paper presented to the Intercultural Communication Division, International Communication Association, Acapulco, Mexico, (May).

Stevenson, Robert L., Richard R. Cole and Donald L. Shaw. (1980). "Patterns of World News Coverage: a Look at the UNESCO Debate on the 'New World Information Order'." Paper presented to the International Communication Division of the Association for Education in Journalism, Boston, Massachusetts, (August).

Stevenson, Robert L. and Kathryn P. White. (1980). "The Cumulative Audience of Network Television News." *Journalism Quarterly. 57*(3), (Autumn), 477–481.

Storey, J. Douglas. (1983). "Television Network News Coverage of the International Affairs Content of Presidential News Conferences." Unpublished Master's Thesis, The University of Texas at Austin, (August).

Television Information Office. (1980). "Public Finds Television News Is Presenting Iran Crisis Objectively, Coping Well with Manipulation Efforts." News Release, Television Information Office, New York, (March 13).

Television News Index and Abstracts. A Guide to the Videotape Collection of the Network Evening News Programs in the Vanderbilt Television News Archive, Joint University Libraries, Nashville, Tennessee, monthly.

Townley, Rod. (1982). "Local News Goes Overseas—And Sometimes Overboard." *TV Guide. 30*(17), (April 24), 10–16.

Townley, Rod. (1981). "The War TV Can't Cover." *TV Guide. 29*(1), (January 3), 5–8.

Tuchman, Gaye. (1978). *Making News: A Study in the Construction of Reality.* New York: The Free Press.

Tunstall, Jeremy. (1977). *The Media Are American.* New York: Columbia University Press.

UNESCO. (1970). *Broadcasting from Space.* UNESCO Reports and Papers on Mass Communication, (60). Paris: UNESCO.

UNESCO. (1976). General Conference Nineteenth Session, Nairobi. 19C/91, Annex I, 3–4.

Weaver, David H. and Judith M. Buddenbaum. (1979). "Newspapers and Television: A Review on Uses and Effects." *ANPA News Research Report 19*, (April 20).

Weaver, David H. and G. Cleveland Wilhoit. (1981). "Foreign News Coverage in Two U.S. Wire Services." *Journal of Communication 31*(2), (Spring), 55–63.

Weaver, Paul H. (1975). "Newspaper News and Television News." In D. Cater and R. Adler, eds., *Television as a Social Force: New Approaches To TV Criticism.* Palo Alto, California: Aspen Institute, 81–94.

Weisman, John. (1983). "Covering International News: The Hazards of Inexperience." *TV Guide 31*(22), (May 28), 2–8.

Westin, Av. (1982). *Newswatch: How TV Decides the News.* New York: Simon and Schuster.

White, Theodore. (1973). *The Making of the President 1972.* New York: Bantam.

Wolzien, Tom. (1980). "Watch Your Loved One Brave Bombs and Bullets." *TV Guide* (January 26). 5–6.

World Bank Atlas. (1980). Washington, D.C.: The World Bank.

"World: The Clouded Window," (1978). Transcript of a television program in the *World* series, produced for PBS by WGBH, Boston, Massachusetts, first broadcast on February 2.

ADDITIONAL REFERENCES

Altheide, David L. and Robert P. Snow. (1979). *Media Logic.* Beverly Hills, California: Sage.

Aronson, James. (1971). *Packaging the News: A Critical Survey of Press, Radio, TV.* New York: International Publishers.

Bagdikian, Ben H. (1971). *The Information Machines: Their Impact on Men and the Media.* New York: Harper and Row.

Barr, Thomas M. (1982). "Decline of the Lease Charge." *COMSAT.* Communications Satellite Corporation Magazine, (8).

Barrett, Marvin. (1978). *Rich News, Poor News: The Sixth Alfred I. duPont Columbia University Survey of Broadcast Journalism.* New York: Thomas Y. Crowell.

Barton, Richard L. and Richard B. Gregg. (1982). "Middle East Conflict as a TV News Scenario: A Formal Analysis." *Journal of Communications 32*(2), (Spring), 172–185.

Becker, Lee B., Maxwell E. McCombs, and Jack M. Mcleod. (1975). "The Development of Political Cognitions." In Steven H. Chaffee, ed., *Political Communication: Issues and Strategies for Research.* Beverly Hills, California: Sage, 21–63.

Bedell, Sally. (1982). "Why TV News Can't Be a Complete View of the World." *New York Times.* (August 8), 1, 20.

Bergsma, Frans. (1978). "News Values in Foreign Affairs on Dutch Television." *Gazette* 24(3), 207–222.

Bogart, Leo. (1977). A Convention Report on the Newspaper Research Project. *The APME Red Book 1977.* New York: The Associated Press.

Buchanan, Patrick. (1978). "The Dangers of Television Diplomacy." *TV Guide 26*(4), (January 28), A5, A6.

Cater, Douglass and Richard Adler, eds. (1975). *Television as a Social Force: New Approaches to TV Criticism.* New York: Praeger.

Cohen, Stanley and Jock Young, eds. (1981). *The Manufacture of News: Social Problems, Deviance and the Mass Media.* revised edition. London: Constable.

Comstock, George and others. (1978). *Television and Human Behavior.* New York:

Columbia University Press.

Diamond, Edwin. (1975). *The Tin Kazoo: Television, Politics and the News.* Cambridge, Massachusetts: The MIT Press.

Diamond, Edwin and Paula Cassidy. (1979). "Arabs vs. Israelis: Has Television Taken Sides?" *TV Guide.* (January 6), 6–10.

Dizard, Wilson P. (1982). *The Coming Information Age: An Overview of Technology, Economics and Politics.* New York: Longman.

Efron, Edith. (1971). *The News Twisters.* Los Angeles, California: Nash.

Elliot, Philip and Peter Golding. (1973). "The News Media and Foreign Affairs." In Robert Boardman and A.J.R. Groom, eds. *The Management of Britain's External Relations.* London: Macmillan.

Fisher, Glen. (1979). *American Communication in a Global Society.* Norwood, New Jersey: Ablex.

Ford, Gerald R. (1981). "How TV Influences a President's Decisions." *TV Guide.* (September 19), 5–8.

Friendly, Fred W. (1977). *The Good Guys, the Bad Guys, and the First Amendment.* New York: Vintage.

Gantz, Walter. (1981). "The Influence of Researcher Methods on Television and Newspaper News Credibility Evaluations." *Journal of Broadcasting* 25(2), (Spring), 155–169.

Gerbner, George, ed. (1977). *Mass Media Policies in Changing Cultures.* New York: Wiley.

Glasgow University Media Group. (1980). *More Bad News.* London: Routledge and Kegan Paul.

Graber, Doris A. (undated). "Approaches to Content Analysis of Television News Programs." Unpublished paper, Department of Political Science, University of Illinois at Chicago Circle.

Graber, Doris A. (1980). *Mass Media and America Politics.* Washington, D.C.: Congressional Quarterly Press.

Groombridge, Brian. (1972). *Television and the People: A Programme for Democratic Participation.* Baltimore, Maryland: Penguin.

Gunter, Barrie. (1981). "Forgetting the News." *Intermedia* 9(5), (September), 41–43.

Hamelink, C. (1976). "An Alternative to News." *Journal of Communication* 25(1), 120–123.

Harms, L. S. and others. (1977). *Right to Communicate: Collected Papers.* Social Sciences and Linguistics Institute, University of Hawaii, Honolulu.

Harris, Phil. (1976). "International News Media Authority and Dependence." *Instant Research on Peace and Violence* 6(4), 148–159.

Hart, Roderick P., Patrick Jerome, and Karen McComb. (1983). "Rhetorical Aspects of Presidential Newscasts." Unpublished manuscript, College of Communication, The University of Texas at Austin, (August).

Hemanus, P. (1976). "Objectivity in News Transmission." *Journal of Communication* 26(4), 102–107.

Hester, Al. (1973). "Theoretical Considerations in Predicting Volume and Direction of International Information Flow." *Gazette* 19(4), 239–247.

Hester, Al. (1974). "The News From Latin America Via a World News Agency." *Gazette* 20(2), 82–98.

Hester, Al. (1978). "Five Years of Foreign News on U.S. Television Evening News-casts." *Gazette 24*(1), 86–95.

Hill, Doug. (1983). "Has Jennings' Flying Circus Paid Off for ABC News?" *TV Guide 31*(9), (February 26), 41–44.

Kaplan, Frank L. (1979). "The Plight of Foreign News in the U.S. Mass Media: An Assessment." *Gazette 25*(4), 233–243.

Kelly, Sean. (1978). *Access Denied: The Politics of Press Censorship.* The Washington Papers, (55). Beverly Hills, California: Sage.

Kinsley, Michael E. (1976). *Outer Space and Inner Sanctums: Government, Business and Satellite Communication.* New York: Wiley.

Kondracke, Morton. (1979). "Eye on the Pentagon—Is TV Telling Us Enough?" *TV Guide.* (August 25), 2–10.

Krippendorff, Klaus. (1980). *Content Analysis: An Introduction to Its Methodology.* Beverly Hills, California: Sage.

Lansipuro, Yrjo, Ibrahim Shahzadeh, and Luke Ang. (1976). *Television News Exchange in Asia.* A case study published by the Asian Mass Communication Research and Information Centre, Singapore, and the Asian Broadcasting Union, (October).

Lawler, George A. (1981). "Comsat Service Bureau, Television Link to the World." *COMSAT.* Communications Satellite Corporation Magazine, (3).

Le Duc, Don R. (1981). "East-West News Flow 'Imbalance': Qualifying the Quantifications." *Journal of Communication 31* (4), (Autumn), 135–141.

Lefever, Ernest W. (1974). *TV and National Defense: An Analysis of CBS News, 1972–1973.* Boston, Virginia: Institute for American Strategy Press.

Legum, Colin and John Cornwall. (1978). *A Free and Balanced Flow: Report of the Twentieth Century Fund Task Force on the International Flow of News.* Lexington, Massachusets: D.C. Heath and Company.

Lent, John A. (1977). "Foreign News in American Media." *Journal of Communication 27*(1), 46–51.

Loory, Stuart H. (1974). "The CIA's Use of the Press—A 'mighty Wurlitzer.'" *Columbia Journalism Review.* (September/October), 9–18.

Marchand, de Montigny. (1981). "The Impact of Information Technology on International Relations." *Intermedia 9*(6), (November), 12–15.

McLure, Robert D. and Thomas E. Patterson. (1976). "Print Versus Network News." *Journal of Communication 26*(2), (Spring), 23–28.

McNulty, T. M. (1975). "Vietnam Specials: Policy and Content." *Journal of Communication 25*(4), 173–180.

McPhail, Thomas L. (1981). *Electronic Colonialism: The Future of International Broadcasting and Communication.* Beverly Hills, California: Sage.

Merrill, John C. and others. (1970). *The Foreign Press: A Survey of the World's Journalism.* Baton Rouge, Louisiana: Louisiana State University Press.

Molotch, Harvey and Marilyn Lester. (1974). "News as Purposive Behavior: On the Strategic Use of Routine Events, Accidents, and Scandals." *American Sociological Review 39* (February), 101–112.

Mueller, Claus. (1973). *The Politics of Communication.* New York: Oxford University Press.

Nessen, Ron. (1981). "Journalism vs. Patriotism: Should TV News Always Tell All?" *TV Guide 29*(26), (June 27), 5–10.

Nichols, John Spicer. (1975). "Increasing Reader Interest in Foreign News by Increasing Foreign News Content in Newspapers: An Experimental Test." *Gazette* *21*, 231–237.

Nnaemeka, Tony and Jim Richstad. (1980). "Structured Relations and Foreign News Flow in the Pacific Region." *Gazette 26*, 235–257.

Nordenstreng, Kaarle, ed. (1974). *Informational Mass Communication*, Helsinki: Tammi Publishers.

Park, Robert E. (1955). *Society Collective Behavior News and Opinion Sociology and Modern Society*. New York: Free Press.

Pollock, John C. (1978). "An Anthropological Approach to Mass Communication Research: The U.S. Press and Political Change in Latin America." *Latin American Research Review 13*(1), 158–172.

Pollock, John C. (1979). "Reporting on Critical Events Abroad: U.S. Journalism and Chile." *Studies in Third World Societies 10* (December), 41–64.

Pollock, John C. (1980). "Becoming a Foreign Correspondent: How Journalists Learn Perspectives on East–West Conflict." Paper presented at the Association for Education in Journalism Annual Conference, Boston, Massachusetts, (August).

Raghavan, Chakravarti. (1976). "A New World Communication and Information Structure." *Development Dialogue 2*, 43–50.

Rarick, Galen, ed. (1971–1972). *News Research for Better Newspapers 6*. Washington, D.C.: American Newspaper Publishers Association Foundation.

Read, William H. (1976). *America's Mass Media Merchants*. Baltimore, Maryland: Johns Hopkins University Press.

Robinson, John P. (1967). "World Affairs Information and Mass Media Exposure." *Journalism Quarterly 44*, 23–31.

Robinson, John P. (1971). "The Audience for National T.V. News Programs." *Public Opinion Quarterly 35*, (Fall), 403–405.

Robinson, John P. and Robert Hefner. (1968). "Perceptual Maps of the World." *Public Opinion Quarterly 32*, 273–280.

Roeh, Itzhak, Elihu Katz, Akiba Cohen, and Barbie Zelizer. (1980). *Almost Mid-Night: Reforming the Late-Night News*. Beverly Hills, California: Sage.

Roshco, Bernard. (1975). *Newsmaking*. Chicago, Illinois: University of Chicago Press.

Rubin, Barry. (1977). *International News and the American Media*. The Washington Papers, (49). Beverly Hills, California: Sage.

Russell, Dick. (1976). "Africa: It Is Drama and Romance—and Now It Has Meaning." *TV Guide*, (December 4), 4–8.

Russett, Bruce and Donald R. Deluca. (1981). "'Don't Tread on Me': Public Opinion and Foreign Policy in the Eighties." *Political Science Quarterly 96*(3), (Fall), 381–399.

Said, Edward W. (1980). "Iran." *Columbia Journalism Review 28*(6), (March/April), 23–33.

Said, Edward W. (1981). *Covering Islam: How the Media and the Experts Determine How We See the Rest of the World*. New York: Pantheon Books.

Sande, Oystein. (1971). "The Perception of Foreign News." *Journal of Peace Research 8*, 221–237.

Schiller, Herbert I. (1970). *Mass Communications and American Empire*. New York: Augustus M. Kelley.

Schiller, Herbert I. (1976). *Communication and Cultural Domination*. White Plains, New York: International Arts and Sciences Press.

Semmel, Andrew K. (1976). "Foreign News in Four U.S. Elite Dailies: Some Comparisons." *Journalism Quarterly* 54(4), 732–736.

Sigal, Leon V. (1973). *Reporters and Officials: The Organization and Politics of Newsmaking*. Lexington, Massachusetts: D.C. Heath.

Signitzer, Benno. (1976) *Regulation of Direct Broadcasting from Satellites: The U.N. Involvement*. New York: Praeger.

Sparkes, Vernone M. and James P. Winter. (1980). "Public Interest in Foreign News." *Gazette 26*, 149–170.

Steinberg, Charles. (1978). "Who Needs Printing? President's Trips Planned for TV News." *TV Guide 26*(15), (April 15), A5, A6.

Stern, Andrew A. (1971). "Impact of Television News." Speech presented to the Radio-Television News Directors Association Meeting, Boston, Massachusetts, (September 29).

Stevenson, Robert L. and Richard R. Cole. (1979). "News Flow Between the Americas: Are We Giving Our Own Hemisphere the Coverage It Deserves?" Paper presented at the 35th annual meeting of the Inter American Press Association, Toronto, Ontario, Canada, (October).

Sussman, Leonard R. (1977). *Mass News Media and the Third World Challenge*. The Washington Papers, (46). Beverly Hills, California: Sage.

U.S. Congress. (1977). Senate Subcommittee on International Operations of the Committee on Foreign Relations. *The Role and Control of International Communications and Information*. Report, 95th Congress, 1st Session. Washington, D.C.: Government Printing Office, (June).

Varis, Tapio. (1973). "European Television Exchanges and Connections With the Rest of the World." *Instant Research on Peace and Violence 1*, 27–43.

Vasquez, Francisco J. and Manny Paraschos. (1980). "Perception of Adequacy of Foreign News Coverage." Unpublished paper, Department of Journalism, University of Arkansas at Little Rock, (September).

de Verneil, Andre J. (1977). "A Correlation Analysis of International Newspaper Coverage and International Economic, Communication and Demographic Relationships.' In Brent D. Ruben, ed., *Communication Yearbook I*. New Brunswick, New Jersey: Transaction Books.

Wald, Richard C. (1977). "Masses and Classes in Communication." *Neiman Reports*. (Summer/Autumn), 27–30.

Weisman, John. (1980). "Stories You Won't See on the Nightly News.' *TV Guide* 28(9), (March 1), 4–8.

Wilhelm, John R. (1972). "The World Press Corps Dwindles: A Fifth World Survey of Foreign Correspondents." Paper read at the International Division Session of the Association for Education in Journalism Convention, Carbondale, Illinois, (August 22).

Williams, Alden. (1975). "Unbiased Study of Television News Bias." *Journal of Communication* 25(4), 190–199.

Wren, Christopher S. (1982). "Cracking China's Great Wall of Silence." *TV Guide*

30(51), (December 18), 5–10.

Yang, Kisuk. (1982). "U.S. Network Coverage of East Asia." M.A. Thesis, The University of Texas at Austin, (August).

Yu, Frederick T. C. and John Luter. (1964). "The Foreign Correspondent and His Work." *Columbia Journalism Review* 3(1), (Spring), 5–12.

Zvi, Dor-Ner. (1977). "How Television Reports Conflicts: Observations of an Unhappy Practitioner." *Nieman Reports.* (Summer/Autumn), 23–26.

Appendix A

Number of Newscasts and Stories Sampled, By Year

Year	ABC		CBS		NBC	
	Newscasts	Stories	Newscasts	Stories	Newscasts	Stories
1972	35	268	35	240	35	259
1973	36	207	35	195	35	209
1974	33	156	36	195	36	202
1975	36	247	36	263	35	239
1976	36	209	36	240	35	179
1977	36	229	36	238	35	199
1978	36	235	36	239	36	257
1979	36	238	36	278	36	261
1980	36	287	36	276	36	234
1981	36	301	36	227	36	247
Totals	357	2377	358	2391	355	2286

Note: The sample consists of 36 randomly selected weeknights per year. During the early years of the period, some broadcasts were pre-empted by local network affiliates in Nashville, Tennessee and therefore are not included in the Vanderbilt Television News Archive.

Appendix B

Content Analysis Units and Coding

The basic unit of analysis for this study is the news item or story. Prior to coding, it was necessary to define the boundaries of a news story, as presented in the *Television News Index and Abstracts*, a process called "unitization" (Holsti, 1969). Fortunately the *Abstracts* are arranged according to news stories and clusters of stories, separated clearly by underlined headings which indicate briefly the topics of stories which follow.

Between the underlined headings, each segment of content begins with a dateline in parentheses, indicating the source or location of the segment. Portions of the newscast read by an anchor correspondent are denoted with "(S)" to show that they originate from the studio or anchor location. Segments which are contributed by network correspondents in New York or Washington, D.C., have "(NY)" or "(DC)" datelines, respectively. Reports originating in other cities of the U.S. or overseas have datelines such as the following: "(KS City,MO)", "(London,England)", "(Jakarta, Indonesia)".

At least one news story follows each underlined heading in the *Television News Index and Abstracts*. The following criteria established clear boundaries for each story unit.

1. Presentation of headlines, previews, or forthcoming stories, usually near the beginning of a newscast, are ignored in the coding process.
2. For purposes of coding, datelines in parentheses are considered to be part of the text of a story.
3. Underlined headings are not considered part of the text of a story.
4. A dateline, followed only by copy read by the anchor correspondent, constitutes one story.
5. A dateline, followed by a string of one or more actualities (film or videotape of actual news events), with script by the anchor correspondent interspersed, is considered a single story.

6. A dateline, followed by a correspondent's report, together with abstract of the anchor correspondent's lead-in and lead-out, is considered a single story.

7. In the case of a serial string of correspondent's reports following one underlined heading, each report with its dateline is considered a separate story. The initial lead-in is part of the first story, the next is part of the second story, and so forth. A lead-out by the anchor correspondent, if present, is considered part of the last correspondent's report.

8. When a network, such as ABC, uses multiple anchor correspondents, the abstracts of material read by different anchors following a single underlined heading are considered separate stories only if the topics of the abstracted material differ.

INFORMATION CODED FOR EACH STORY UNIT

In the process of coding content data, coders searched each news story for the following recording units:

1. Mentions of nations or dependent territories.
2. Mentions of certain international organizations.
3. Use of film or videotape.
4. Origin of film or videotape: domestic or foreign.
5. Nation of origin for video reports originating overseas. (This information was only gathered for six years, 1976–1981, in the case of NBC.)
6. Story theme. (For NBC and ABC this information was coded only for the 1976–1981 period.)
7. Length of story in seconds. (For all three networks, this information was coded only for 1976–1981 period.) Data on story length were recorded directly from the time notations that appear in the left hand margin of the *Television News Index and Abstracts*. If time was indicated only for a cluster of stories, story length was recorded as the total cluster time divided by the number of stories in the cluster.
8. Rank of story in the newscast. (For all networks, coded only for 1976–1981 period.)

In addition to the above recording units, each news story in the data file was identified by network, and by month, date, year, and weekday of broadcast. Also, the total number of stories in the newscast, both international and domestic, was noted for each international news item.

CODING REFERENCES TO NATIONS, TERRITORIES, AND ORGANIZATIONS

As a general rule, only *manifest* mentions of nations, dependent territories, and international organizations such as the United Nations were coded. *Latent* content was ignored. Only one mention was coded per news story, regardless of the actual number of references. This created a series of dichotomous variables indicating whether a nation, territory or organization was mentioned at least once in any news story.

Mentions of prominent place names, such as capital cities of nations were coded as if the country itself were mentioned. Likewise, references to Premiers, Foreign Ministers, Kings, Queens or comparable governmental officials were coded as references to their respective nations. If a country was used as an adjective, it may or may not have been coded. For example, the following would be considered references to nations: British Petroleum, Irish Republican Army, Vietcong. However, if a nation is mentioned in a context that clearly indicates a reference to something other than the country itself, it is not coded. Some examples are: Sea of Japan, Indian Ocean, Mexican-American, Air France, Singapore Airlines, Greek Mythology.

The only exception to the above coding rules involves the United States. Because network television is broadcast to American audiences, the United States is often not included as part of the manifest content in the *Television News Index and Abstracts*. For example, it was often not necessary to identify Secretary of State Henry Kissinger as "U.S. Secretary of State Kissinger." Therefore, the following general rule was used for the United States: Whenever a news story involved the U.S. and another nation or nations, the United States was coded, even if not included in the manifest content of the abstracts. Inclusion of the U.S. was implied by mention of prominent U.S. institutions, leaders, and place names.

In the case of general references to divided nations, the following convention was used: Viet Nam referred to South Vietnam (prior to unification in 1976), Korea was coded as Republic of Korea (South Korea), Berlin was coded as West Germany, Germany was coded as West Germany, and so forth.

CODING OF STORY THEMES

Although the source of data was in the form of abstracts, rather than a verbatim text of each newscast, thematic content was coded for each news item. Coders noted only the *major theme* found in the abstract of each story. Importance of the theme was indicated by the number of sentences or words devoted to it in the abstract. If a story appeared to contain two equally important themes, only the first to appear in the abstract was coded. The specific themes coded are listed in the coder guide which follows.

CODER GUIDE

For each international news story in the sample, coders noted information on the following list of variables, as indicated. The data was recorded on standard 80-column worksheets for later data entry. Each case (news item) required four cards or card images.

1. Network (ABC = 1;CBS = 2;NBC = 3)
2. Month (January = 1;February = 2..December = 12)
3. Date (01 through 31)
4. Year (72 through 81)
5. Day (Monday = 1;Tuesday = 2...Friday = 5)
6. Video (Video = 1;Nonvideo = 2)
7. Location (Foreign Video = 1;Domestic Video = 2)
8. Source (Albania = 1,Andorra = 2,Austria = 3,Belgium = 4... Zambia = 221,Washington, D.C. = 400,New York City = 401, Other U.S. = 402) A different number code was used for each nation and territory, and for U.S. locations as indicated.
9. Theme (each news story was placed in one of the following nine thematic categories)
    a. Unrest and Dissent
    b. War,Terrorism,Crime
    c. Coups and Assassinations
    d. Natural and Man-Caused Disasters
    e. Political–Military Affairs
    f. Economics,Trade
    g. Environment
    h. Technology–Science
    i. Human Interest
10. Rank (1,2,3,etc., for the rank of each story in the newscast)
11. Length in Seconds (020,030,180,etc.)
12. Total (Total number of stories in newscast, 10,11,12,etc.)

The remaining variables were coded using a complete list of all the nations and dependent territories in the world, and the United Nations. Each variable takes a value of "1" if mentioned in the news item and a value of "0" if not mentioned.

Appendix C

Reliability of the Television News Index and Abstracts as a Measure of Coverage of Nations, 1972-1976

Country	Reliability[a]	Number of Mentions of Nations on Audio Tapes
1. United States	.89	130
2. South Vietnam	.94	48
3. USSR	.81	33
4. North Vietnam	.87	33
5. Israel	.94	19
6. France	.87	13
7. Great Britain	.87	18
8. Cambodia	.84	14
9. Egypt	.81	9
10. People's Republic of China	.79	11
11. West Germany	.88	9
12. Japan	.91	13
13. Syria	1.00	9
14. Lebanon	1.00	6
15. Northern Ireland	.78	8
16. Switzerland	.89	5
17. Turkey	.89	5
18. The Philippines	.89	5
19. Laos	.65	7
20. Pakistan	.91	6
21. Cuba	.94	9

[a] As indicated by a Pearson zero-order correlation coefficient between mentions on audio tapes and mentions in the *Television News Index and Abstracts.*

Nations and Territories of the World

The following listing of nations and territories of the world is grouped by geographical region. Information on economic category, (D = Developed Market Economy, LD = Developing Market Economy, S = Centrally Planned Economy), presence of AP or UPI correspondents (P = Associated Press or United Press International), and possession of an Intelsat earth station (I = Intelsat Earth Station). If AP, UPI or Intelsat were present for less than the entire 1972–1981 decade, the years will be indicated with the listing.

North America

1. Nations

   United States of America, D, P, I
   Canada, D,P,I

2. Territories

• St. Pierre and Miquelon,LD

Latin America

1. Nations

   Argentina,LD,P,I
   Bahama Islands,LD
   Barbados,LD,I 73-81
   Bolivia,LD,P,I 79-81
   Brazil,LD,P,I

Belize,LD,I 79-81
Chile,LD,P,I
Colombia,LD,P,I
Costa Rica,LD
Cuba,S,I 80-81
Dominica,LD
The Dominican Republic,LD,P,I 75-81
Ecuador,LD,P 77-81,I 73-81
El Salvador,LD,P 79-81,I 78-81
Grenada,LD
Guatemala,LD,P 79-81,I 80-81
Guyana,LD,I 80-81
Haiti,LD,I 77-81
Honduras,LD
Jamaica,LD,P 79-81,I
Mexico,LD,P,I
Netherlands Antilles,LD,P 79-81,I 79-81
Nicaragua,LD,P 79-81,I 73-81
Panama,LD,P 79-81,I
Paraguay,LD,P 77-81,I 78-81
Peru,LD,P,I
St. Lucia,LD
St. Vincent and the Grenadines,LD
Suriname,LD,I 79-81
Trinidad and Tobago,LD,I
Uruguay,LD,P,I 80-81
Venezuela,LD,P,I

2. Dependent Territories

Bermuda,LD
British Virgin Islands, LD
Cayman Islands, LD
Guadeloupe,LD
French Guiana,LD
Falkland Islands,LD
Martinique,LD
Montserrat,LD
Panama Canal Zone,LD
Puerto Rico,LD
U.S. Virgin Islands,LD

Africa

1. Nations

    Angola,LD,I 75-81
    Benin,LD (formerly Dahomey)
    Botswana,LD,I 81
    Burundi,LD,I 81
    Cameroon,LD,I 74-81
    Central African Republic,LD
    Chad,LD
    Djibouti,LD,I 81 (formerly Afars-Issas)
    Congo,People's Republic of,LD,I 78-81
    Equatorial Guinea,LD
    Ethiopia,LD,I 79-81
    Gabon,LD,I 74-81
    The Gambia,LD,I 79-81
    Ghana,LD,P 79-81
    Guinea,LD,I 81
    Guinea-Bissau,LD
    Ivory Coast,LD,I 73-81
    Kenya,LD,P,I
    Lesotho,LD
    Liberia,LD,P 79-81,I 77-81
    Madagascar,LD,I 78-81
    Malawi,LD,I 77-81
    Mali,LD,I 77-81
    Mauritania,LD
    Mozambique,LD,I 75-81
    Namibia,LD
    Niger,LD,I 78-81
    Nigeria,LD,P,I
    Zimbabwe,LD,P 79-81 (formerly Rhodesia)
    Rwanda,LD
    Sao Tome and Principe,LD
    Senegal,LD,P 79-81,I 73-81
    Sierra Leone,LD,I 78-81
    Somalia,LD,I 79-81
    South Africa,D,P,I 76-81
    Sudan,LD,I 75-81
    Swaziland,LD
    Tanzania,LD,I
    Togo,LD,I 78-81

Upper Volta,LD,I 78-81
Uganda,LD,P 79-81,I
Zaire,LD,I
Zambia,LD,P 79-81,I 75-81

2. Dependent Territories

St. Helena,LD
Ascension,LD
Tristan da Cunha,LD
Cueta-Melilla,LD (Spanish Sahara)

Eastern Europe and the USSR

1. Nations

Albania,S
Bulgaria,S
Czechoslovakia,S
East Germany,S
Hungary,S
Poland,S,P
Romania,S,P 79-81,I 77-81
The USSR,S,P,I 75-81
Yugoslavia,S,P,I 74-81

2. Dependent Territories

(None)

Western Europe

1. Nations

Andorra,D
Austria,D,P,I 80-81
Belgium,D,P,I 73-81
Cyprus,LD,I 80-81
Denmark, D,P,I
Finland,D,P,I
France,D,P,I
West Germany,D,P,I
Gibraltar,D,I 80-81

Greece,D,P,I 73-81
Iceland,D,I 81
Ireland,D
Italy,D,P,I
Liechtenstein,D
Luxembourg,D,P
Malta,D
Monaco,LD
The Netherlands,D,P,I 74-81
Norway,D,P,I
Portugal,D,P,I 75-81
San Marino,LD
Spain,D,P,I
Sweden,D,P,I
Switzerland,D,P,I 74-81
Turkey,LD,P,I 78-81
United Kingdom,D,P,I
Vatican City (special status)

2. Dependent Territories

Faeroe Islands, D
Greenland,LD
Northern Ireland,D
Isle of Man,LD
Channel Islands,LD

The Middle East and North Africa

1. Nations

Algeria,LD,P 79-81,I 76-81
Bahrain,LD,P 79-81,I
Egypt,LD,P,I 75-81
Iran,LD,P,I
Iraq,LD,I 76-81
Israel,D,P,I 73-81
Jordan,LD,P 79-81,I
Kuwait,LD,I
Lebanon,LD,P,I
Libya,LD,I 79-81
Morocco,LD,I
Oman,LD,I 77-81

Qatar,LD,I 76-81
Saudi Arabia,LD,I 76-81
Syria,LD,I 78-81
Tunisia,LD,P 79-81
United Arab Emirates,LD,I 76-81
Yemen Arab Republic,LD,I 77-81
Yemen,People's Democratic Republic,LD,I 81

2. Dependent Territories

(None)

Asia and the Pacific

1. Nations

Afghanistan,LD
Australia,D,P,I
Bangladesh,LD,I 76-81
Bhutan,LD
Brunei,LD,I 80-81
Burma,LD,P 79-81,I 80-81
China,People's Republic,LD,P 79-81,I
The Comoros,LD,(independence from France,1975)
Fiji,LD,I 77-81
India,LD,P,I
Indonesia,LD,P,I
Japan,D,P,I
Kampuchea,LD,P 72-75 (formerly Cambodia)
Kiribati,LD, (Gilbert Is. before 1979)
Korea,Democratic People's Republic,S
Korea,Republic of,LD,P,I
Laos,LD
Malaysia,LD,P,I
Maldives,LD,I 78-81
Mauritius,LD
Mongolia,S
Nepal,LD,P 79-81
New Zealand,D,I
Pakistan,LD,P,I 73-81
Papua New Guinea,LD (Independence from Australia,1975)
The Philippines,LD,P,I
Seychelles,LD

Singapore,LD,P,I
Sri Lanka,LD,P 79-81,I 76-81
Solomon Islands,LD
Taiwan,LD,P,I
Thailand,LD,P,I
Tonga,LD,I 79-81
Tuvalu,LD (Ellice Islands until 1975)
Vanuatu,Republic of,LD,I 80-81 (New Hebrides until 1980)
Vietnam,S,P 72-75 (unification in 1976)
Western Samoa,LD,I 81

2. Territories

American Samoa,LD
Cocos Islands,LD (Australia)
Cook Islands,LD (New Zealand)
Christmas Island,LD (Australia)
French Polynesia,LD
French Southern and Antarctic Territories,LD
Guam,LD (U.S.)
Hong Kong,D,P,I
Macao,LD
New Caledonia,LD (France)
Niue,LD (New Zealand)
Norfolk Island,LD (Australia)
Pitcairn Islands,LD (Britain)
Portugese Timor,LD
Reunion,LD
Ross Dependency,LD (New Zealand)
Tokelau,LD (New Zealand)
Trust Territory of the Pacific Islands,LD (U.S.)
Turks and Caicos Islands,LD (Britain)
Wallis and Futuna Islands,LD (France)

Sources of Information

Nations were classified as developed, developing, or socialist based on the system used in the *Statistical Yearbook 1978*, published by the United Nations. A number of the developing nations, notably the oil-exporting countries, have 1978 per capita gross national products of more than $3,000 according to the *World Bank Atlas*, Washington, D.C., World Bank, 1980.

Information on the presence of AP or UPI reporters came from two sources. One was the survey by Kliesch (1975) published in the membership directory of the Overseas Press Club of America. The other was *The Europa Yearbok* for 1981.

Data concerning the nations served by Intelsat directly through an earth station or indirectly through terrestrial connections with a nearby earth station were gathered from the following sources.

COMSAT Guide to the Intelsat, Marisat and Comstar Satellite Systems. (1979, 1980). Washington, D.C.: Comsat Office of Public Information.

COMSAT Guide to the Intelsat Satellite System (1982). Washington, D.C.: Office of Corporate Affairs, Communications Satellite Corporation.

COMSAT Pocket Guide to the Global Satellite System, 1978. (1978). Washington, D.C.: Comsat Office of Public Information.

The INTELSAT Global System Map. (1982). Washington, D.C.: Communications Satellite Corporation, Office of Corporate Affairs, (January).

INTELSAT 1981 Annual Report. (1981). Washington, D.C.: International Telecommunications Satellite Organization.

Listing of Existing and Proposed Earth Stations Through 1981. (1977). Internal Memorandum, COMSAT and COMSAT General, (May 20).

Appendix E

Bivariate Regression Analyses with Time as the Independent Variable and the Three Major Story Formats as Dependent Variables, All Networks, 1972-1981

Dependent Variable	F	Significance	Simple R	R^2	Degrees of Freedom	
					Regression	Residual
1. Anchor Reports	155.38	.000	$-.15$	.02	1	7052
2. Domestic Video Reports	55.67	.000	.09	.007	1	7052
3. Foreign Video Reports	36.54	.000	.07	.005	1	7052

Note: The independent variable, time, takes a value of 1 through 10, with each number representing a year during the 1972–1981 decade.

Fifty Most Frequently Mentioned Nations, by Network (ABC, CBS, NBC) and by Year (1972-1981)

Tables F.1, F.2, and F.3 are found on the following pages. Because of tie ranks and cross-network differences in ranks, each table includes more than fifty nations. The N entries at the end of each table represent the total number of international news stories in the sample, by year and for the entire decade. The N columns show the number of sampled stories which mention each nation in the table, respectively.

Table F.1
ABC Evening News Coverage of 50 Countries, 1972-1981

Rank Nation	1972%	1973%	1974%	1975%	1976%	1977%	1978%	1979%	1980%	1981%	Total %	N
1. United States	63.8	62.8	51.9	61.9	52.2	60.3	64.7	60.5	53.3	41.2	57.0	1355
2. USSR	13.4	8.7	21.8	16.2	18.7	16.2	23.0	13.9	21.3	15.3	16.7	398
3. Israel	9.0	17.9	18.0	15.4	7.7	17.5	21.7	16.0	9.4	13.6	14.3	340
4. Britain	7.5	7.7	7.7	7.7	10.5	13.5	14.5	9.7	5.9	13.3	9.8	234
5. South Vietnam	40.3	20.8	5.1	17.4	5.7	0	0	0	0	0	9.1	215
6. Iran	1.0	1.0	0	0.8	0.5	9.6	0.9	27.7	35.9	9.3	8.7	206
7. Egypt	1.1	7.7	9.0	10.1	1.4	6.1	14.0	12.2	5.6	7.6	7.7	184
8. North Vietnam	30.2	15.5	0.6	3.2	0.5	9.6	4.3	8.0	1.7	2.3	7.5	178
9. France	11.2	7.3	5.1	3.6	6.2	9.6	6.4	4.6	3.5	4.7	6.2	147
10. China (P.R.)	11.6	4.4	3.2	4.5	8.6	2.2	3.4	8.4	4.5	2.3	5.3	127
11. Lebanon	1.9	2.4	3.9	5.3	12.9	4.4	6.4	0.8	1.1	4.3	4.2	99
12. West Germany	5.2	2.4	1.3	1.6	1.9	9.2	4.7	3.4	4.2	5.7	4.1	98
13. Japan	4.9	7.3	3.9	1.2	5.7	4.4	3.0	5.5	3.5	2.7	4.1	97
14. Syria	1.5	5.8	10.3	4.5	4.8	4.4	1.7	0.4	1.4	3.7	3.5	83
15. Cuba	0.8	1.0	1.3	1.2	6.7	4.8	3.8	5.5	3.1	3.3	3.2	75
16. Poland	1.5	0.5	0	0	0.5	0	1.7	0.8	6.6	14.0	3.1	73
17. Saudi Arabia	0	3.4	1.9	1.6	1.9	3.5	2.1	4.2	1.1	8.0	2.9	68
18. Italy	1.1	1.5	1.9	1.6	4.3	4.4	5.1	4.2	1.7	2.7	2.8	67
18. Cambodia	1.5	13.5	1.9	6.9	0.5	0.9	0.4	4.2	0	0.3	2.8	67
19. Afghanistan	0	0	0	0	0.5	0	0	2.1	15.3	1.3	2.3	54
20. Rhodesia(Zimbabwe)	0	0	0	0	4.8	3.9	6.0	4.2	2.4	1.0	2.2	53
21. South Africa	0	0	0	0.4	6.7	8.3	2.1	1.7	0.4	2.3	2.2	51
22. Northern Ireland	4.1	2.4	3.2	1.2	2.9	0	0	2.1	0.7	4.0	2.1	49
23. Canada	0.4	1.9	3.2	1.2	3.8	3.1	2.6	0	2.1	1.3	1.9	44
24. Iraq	0.4	1.0	0	0.4	1.4	0.4	0.4	1.7	5.9	4.3	1.8	43
25. Turkey	0	0	10.9	2.4	1.9	0	1.3	1.7	0.4	1.0	1.6	38
25. Jordan	0.8	0	1.9	2.8	0.5	4.4	3.0	2.5	0.4	0.3	1.6	38
26. Switzerland	0.4	0.5	1.3	1.2	1.4	4.8	1.3	2.5	0.7	0.7	1.4	34
26. Libya	0	1.9	0.6	0	1.4	0.9	0	0.8	3.5	4.0	1.4	34
27. Mexico	0.8	0.5	1.9	0	2.9	0.9	0.9	3.8	1.1	1.3	1.4	32
28. South Korea	2.2	0	0	1.2	1.0	5.2	1.7	0.4	0.4	0.7	1.3	31
29. India	1.9	1.0	1.3	1.6	0	3.5	0.9	0.4	1.7	0.3	1.3	30

Table F.1 (continued)

Rank	Nation	1972%	1973%	1974%	1975%	1976%	1977%	1978%	1979%	1980%	1981%	Total %	N
30.	Spain	0.8	0	1.3	2.4	2.4	3.1	0.4	0.4	0.4	1.0	1.2	28
30.	Pakistan	1.9	0.5	0.6	0	0.5	0.9	0	1.3	2.4	2.7	1.2	28
30.	Panama	0	0.5	0	0	0	3.1	4.3	1.3	2.1	0.3	1.2	28
31.	Cyprus	0	0.5	10.3	1.6	0.5	0	1.7	0	0	0	1.1	26
31.	Greece	0	1.9	6.4	1.2	1.4	1.3	0	0	0.7	0.3	1.1	26
32.	Philippines	2.6	1.5	1.3	1.6	1.4	0	0	1.3	0.4	0.3	1.0	24
33.	Thailand	0.8	0.5	0	2.4	2.4	0.4	0	3.4	0	0	1.0	23
33.	The Vatican	0.4	0	0	0.8	0	0	3.4	0.4	1.7	2.0	1.0	23
34.	Netherlands	0	0.5	1.3	2.8	1.0	1.8	0.9	0	0	1.3	0.9	22
34.	Algeria	0	1.5	1.3	0.4	0	0.4	0	0.8	4.2	0.3	0.9	22
34.	Angola	0	0	0	0.8	6.7	0.9	0.9	0	0	0.7	0.9	22
34.	Laos	1.1	4.4	1.3	2.0	0.5	0	0.5	0.4	0	0	0.9	22
34.	Uganda	0	0	0	1.6	2.4	3.5	0.4	1.7	0	0	0.9	22
35.	Portugal	0	0	1.9	4.9	1.0	0.9	0	0.4	0	0	0.8	20
36.	Austria	0.4	0	0.6	1.6	0.5	0.9	1.3	1.3	1.4	0	0.8	19
37.	Argentina	0	1.0	1.3	2.4	2.4	0.4	0.9	0	0	0	0.8	18
38.	Nicaragua	0.4	0.5	0	0	0	0	2.6	1.7	0.7	0.7	0.7	16
38.	Chile	0.4	1.0	1.3	0.9	1.0	0.4	1.3	0.4	0.4	0.3	0.7	16
38.	Sweden	1.1	0.5	0.6	0.8	1.4	0.9	0.9	0	0	0.7	0.7	16
38.	East Germany	0.4	1.0	0.6	0	0.5	0.4	0.4	0.8	1.7	0.7	0.7	16
	N =	268	207	156	247	209	229	235	238	287	301	2377	

Table F.2
CBS Evening News Coverage of 50 Countries, 1972-1981

Rank Nation	1972%	1973%	1974%	1975%	1976%	1977%	1978%	1979%	1980%	1981%	Total %	N
1. United States	72.5	65.6	54.9	61.6	52.9	63.5	67.0	60.8	56.2	49.8	60.5	1446
2. USSR	17.5	11.3	19.0	15.6	14.6	13.9	26.8	13.3	20.3	18.1	17.1	408
3. Israel	4.2	16.9	14.9	10.3	10.0	18.5	16.3	15.8	14.5	13.2	13.4	320
4. Britain	7.5	7.7	15.9	9.9	12.9	8.8	8.8	7.9	9.1	11.5	9.9	236
5. South Vietnam	40.0	22.1	2.6	19.8	4.6	0	0	0	0	0	8.7	207
6. Iran	1.7	1.0	2.6	0.8	0.8	0.4	2.5	27.3	29.4	10.1	8.5	202
7. Egypt	1.7	8.7	5.6	4.6	4.6	9.7	13.8	14.0	6.2	7.1	7.7	183
8. France	12.9	9.7	8.2	5.3	7.5	11.3	4.6	4.7	2.2	7.5	7.2	172
9. North Vietnam	28.8	15.4	0	2.7	0.4	5.9	5.4	9.0	1.5	3.1	7.1	170
10. China (P.R. of)	8.8	2.6	3.6	4.2	7.1	1.7	4.6	7.2	3.6	2.2	4.6	111
11. West Germany	3.3	5.1	5.1	4.2	4.6	8.4	5.0	2.5	2.5	3.5	4.4	104
12. Lebanon	0.8	2.1	2.6	4.9	13.3	3.8	4.2	1.8	1.5	4.0	3.9	93
13. Saudi Arabia	0	2.1	2.6	1.9	3.3	3.4	3.8	4.7	2.5	10.6	3.5	83
13. Syria	0	6.2	7.7	2.3	7.1	5.0	3.4	1.4	1.1	2.6	3.5	83
14. Japan	3.8	5.1	4.1	2.7	3.3	2.5	2.5	4.3	4.0	2.2	3.4	82
15. Cuba	1.3	1.5	1.5	1.9	2.5	5.0	4.6	3.6	5.4	4.0	3.2	77
16. Cambodia(Kampuchea)	2.9	13.9	1.0	8.4	0	0.8	0.8	2.9	1.1	0	3.1	73
17. Poland	0.8	1.0	0	0.8	0.8	0.4	1.7	1.8	6.9	14.1	2.9	69
18. South Africa	0	0	1.0	1.5	7.9	7.1	3.4	2.9	1.1	1.8	2.7	65
19. Canada	1.7	1.0	2.1	1.1	3.3	3.8	3.8	2.5	2.5	3.5	2.6	61
19. Switzerland	1.7	1.0	3.1	1.9	2.9	8.0	2.5	2.9	1.1	0.4	2.6	61
20. Italy	0.8	1.0	2.6	3.0	2.5	4.2	4.6	1.8	1.5	1.3	2.3	56
21. Rhodesia(Zimbabwe)	0.4	0	0	0	5.4	4.2	5.9	3.6	1.8	0	2.2	53
22. South Korea	0.8	1.0	1.0	2.7	2.1	6.7	3.4	0.7	0.4	0.4	1.9	46
23. Mexico	1.3	0	0.5	0.8	2.9	2.5	1.3	3.6	1.5	4.0	1.9	45
24. Jordan	0.8	1.5	2.6	1.1	0.8	5.5	2.9	1.4	0.7	1.3	1.8	44
25. Libya	0	2.6	0.5	0.4	1.3	0.8	0.4	1.8	3.6	5.7	1.7	41
26. Northern Ireland	3.3	1.0	3.6	0.4	2.5	0.8	0	0.7	0.4	3.5	1.6	37
26. Spain	1.7	0.5	0.5	2.3	3.8	2.9	0.4	1.8	0.7	0.4	1.6	37
27. Turkey	0	0	6.2	1.9	2.5	0.4	2.1	1.1	0.7	0.4	1.5	35
28. Panama	0	0.5	0	0	2.5	2.9	5.0	0.7	2.9	0	1.4	34
29. Afghanistan	0	0	0	0	0	0	0	0.7	9.8	1.8	1.4	33

Table F.2 (continued)

Rank	Nation	1972%	1973%	1974%	1975%	1976%	1977%	1978%	1979%	1980%	1981%	Total %	N
30.	Iraq	0	1.0	1.0	0.4	2.1	0.4	0.8	1.8	2.2	3.1	1.3	31
30.	Greece	0	1.0	4.6	1.9	1.3	1.3	2.1	0.4	0.8	0.4	1.3	31
31.	India	1.7	0	1.0	0.8	0.8	1.3	2.5	0.4	2.2	0.4	1.1	27
32.	Portugal	0	0	2.6	5.7	2.5	0	0	0	0	0	1.1	26
33.	Cyprus	0	0	7.7	0.8	0	0	2.5	0.4	0	0	1.0	24
33.	Sweden	1.3	1.0	2.1	0.4	1.3	0	0.8	0.7	1.8	0.9	1.0	24
34.	Thailand	1.7	1.0	0.5	2.3	0	0.4	0.4	2.9	0	0	1.0	23
34.	Chile	1.3	4.1	0.5	1.1	0.4	0.8	1.3	0	0.7	0	1.0	23
35.	Angola	0	0	0	1.5	5.0	0.4	1.7	0	0	0	0.9	21
35.	Algeria	0.4	1.5	0.5	1.1	1.3	0.4	0	1.1	2.2	0	0.9	21
35.	Pakistan	0.8	0	0.5	0.8	0	0	0	1.1	2.9	2.2	0.9	21
35.	Uganda	0	0	0	1.5	1.3	2.5	0.8	1.4	0.4	0.4	0.9	21
35.	The Vatican	0	0	0.5	0.4	0	0	2.9	1.1	1.1	2.6	0.9	21
35.	Yugoslavia	0.4	0	0	0.8	1.7	0.8	2.1	0.4	1.8	0.4	0.9	21
36.	The Philippines	1.3	0	0	1.9	0.8	0.8	0	0.7	0	0	0.8	20
37.	Argentina	0	3.1	1.5	2.3	0.8	0.8	0.4	0	0.7	0	0.8	19
37.	The Netherlands	0	1.5	1.0	0.8	2.1	1.7	0	0.7	0.7	0	0.8	19
38.	El Salvador	0	0	0	0	0	0.4	0	1.4	1.1	4.4	0.8	18
38.	East Germany	0.4	0.5	1.0	0.4	0.4	1.3	0.8	0.4	1.1	1.3	0.8	18
	N =	240	195	195	263	240	238	239	278	276	227		2391

Table F.3
NBC News Coverage of 50 Countries, 1972–1981

Rank	Nation	1972%	1973%	1974%	1975%	1976%	1977%	1978%	1979%	1980%	1981%	Total%	N
1.	United States	65.6	64.6	49.5	59.4	53.1	68.8	56.0	59.0	58.6	52.2	58.8	1343
2.	USSR	13.9	10.1	20.3	13.4	13.4	15.1	19.5	14.6	23.5	17.8	16.2	371
3.	Israel	4.3	17.2	17.8	11.7	7.3	16.1	20.2	16.1	13.3	11.7	13.6	310
4.	South Vietnam	40.9	21.1	2.5	16.3	0	0	0	0	0	0	9.0	205
5.	Britain	7.0	5.7	12.9	6.7	11.2	13.6	7.0	4.6	6.4	14.6	8.8	200
6.	Egypt	2.7	9.1	6.9	10.0	3.9	5.0	13.2	13.0	5.1	8.9	8.0	183
7.	North Vietnam	31.3	14.4	0	5.4	0	6.0	3.9	8.8	2.1	1.6	7.8	178
8.	Iran	1.2	1.4	0.5	0.4	2.2	1.5	1.6	23.8	30.3	6.9	7.4	169
9.	France	13.9	8.6	6.9	4.6	6.7	6.5	5.8	3.8	2.6	3.6	6.3	144
10.	China (P.R. of)	8.1	3.4	4.0	3.4	7.8	1.0	3.9	9.2	3.0	2.0	4.6	106
11.	West Germany	3.9	3.8	3.5	3.4	5.6	5.5	5.5	2.7	2.6	4.1	4.0	91
12.	Saudi Arabia	0	3.4	4.0	1.7	0.6	3.5	4.3	4.6	1.7	8.5	3.3	75
13.	Syria	1.2	5.3	10.4	3.4	5.0	2.5	2.3	0.8	1.3	2.4	3.2	74
14.	Cuba	1.9	1.0	1.5	2.5	5.0	3.5	3.9	3.5	6.0	3.2	3.2	73
15.	Japan	4.6	4.8	5.0	1.7	2.8	2.5	1.2	1.9	3.0	4.5	3.2	72
16.	Italy	1.2	1.4	2.0	2.9	2.8	4.0	6.2	3.5	3.0	3.6	3.1	71
17.	Poland	1.5	1.9		0.8	1.1	0.5	2.0	1.2	6.4	13.8	3.0	69
17.	Lebanon	0.1	1.9	5.0	4.6	10.6	1.0	3.9	1.2	1.7	1.6	3.0	69
18.	Cambodia	3.1	14.8	0.5	6.7	0	0.5	0.8	3.1	0	0.4	3.0	68
19.	Canada	1.2	1.9	1.5	1.7	3.4	3.5	3.1	2.3	3.9	1.2	2.3	53
20.	South Korea	1.5	1.0	0.5	2.5	1.7	5.5	3.5	2.3	1.3	0.8	2.1	47
21.	Mexico	0.4	0.5	1.0	0.4	1.7	3.5	2.7	3.1	1.3	2.8	1.8	40
22.	South Africa	0.4	0		0	6.7	6.0	2.0	2.3	0.4	0.8	1.7	39
23.	Afghanistan	0	0	0	0	0	0	0.4	1.2	12.0	1.2	1.5	35
23.	Northern Ireland	4.6	2.4	2.5	0.8	0.6	0	0	0	0.9	3.2	1.5	35
24.	Switzerland	0.8	0.5	3.5	1.7	1.1	3.5	0.8	2.3	0.4	0.4	1.4	33
24.	Thailand	1.5	1.9	0.5	4.2	1.7	0.5	0	3.8	0	0	1.4	33
25.	Rhodesia(Zimbabwe)	0	0	0	0	4.5	1.5	3.5	3.1	0.9	0	1.3	30
25.	Greece	0	1.0	6.9	1.3	1.7	1.5	0.8	0	0.4	0.8	1.3	30
26.	Spain	0	0.5	1.5	2.9	3.9	3.0	0	1.5	0.8	0.8	1.3	30
26.	Libya	0	1.9	0.5	0	0.6	0	0.4	1.9	5.6	1.6	1.3	29
26.	Jordan	0.4	1.0	3.5	1.3	0.6	2.0	1.6	1.9	0.4	0.4	1.3	29

Table F.3 (continued)

Rank	Nation	1972%	1973%	1974%	1975%	1976%	1977%	1978%	1979%	1980%	1981%	Total %	N
27.	Turkey	0	0	6.4	1.3	2.8	0	1.6	0.4	0.4	0.4	1.2	28
27.	Iraq	0.8	0	0.5		0.6	0	1.2	2.3	2.6	3.6	1.2	28
28.	India	2.7	0.5	0.5	1.3	0	2.0	1.6	0.4	1.3	0.4	1.1	25
29.	The Vatican	0.4	0	0.5	0.8	0	0	3.1	1.2	0.9	2.8	1.1	24
30.	Portugal	0	0	1.5	5.4	2.2	0	0.4	0	0.4	0	1.0	22
30.	Laos	1.9	3.8	0.5	1.7	0	0	0.4	1.2	0	0	1.0	22
31.	Cyprus	0.4	0	7.4	0	0	0	2.0	0	0	0	0.9	21
32.	Uganda	0.8	0	0	1.3	1.7	3.5	0.8	1.2	0	0	0.9	20
33.	Pakistan	2.3	0	0.5	0	0	0	0.4	1.2	1.3	2.0	0.8	19
33.	East Germany	0	0.5	0.5	0	0.6	1.0	0.8	1.2	1.7	2.0	0.8	19
33.	Belgium	1.5	0	1.0	1.7	1.1	1.0	0.8	0	0.9	0.4	0.8	19
34.	The Philippines	1.5	2.9	0	1.3	1.1	0.5	0.8	0	0	0	0.8	18
34.	Taiwan	2.7	0	0	0.4	0	0	1.6	1.5	0.9	0	0.8	18
34.	Angola	0	0	0	1.7	5.6	1.0	0.8	0	0	0	0.8	18
34.	Chile	2.3	1.0	2.5	0.4	0.6	0.5	0.4	0	0.4	0	0.8	18
35.	The Netherlands	0.8	1.4	1.5	1.3	0.6	2.5	0	0	0	0	0.7	17
35.	El Salvador	0	0	0	0	0	0	0	0.4	1.7	4.9	0.7	17
35.	Panama	0	0	0	0	1.1	2.0	3.1	0	0.9	0.4	0.7	17
36.	Algeria	0.4	1.0	1.0	0	0.6	0.5	0	1.2	2.1	0.4	0.7	16
	N =	259	209	202	239	179	199	257	261	234	247	2286	

Author Index

Italic page numbers indicate bibliographic citations.

A

Adams, W. C., 20, 21, 31, 32, 131, 144, 149, *152*
Adoni, H., 11, *155*
Aggarwala, N. K., 94, *152*
Almaney, A., 35, *152*
Anderson, M. H., 17, 19, 20, *157*

B

Batscha, R. M., 16, 26, 27, 29, 37, 102, 115, 116, 134, 144, *152*
Bogart, L., 8, 10, 11, 12, 49, *152*
Boorstin, D. J., 26, *153*
Bower, R. T., 9, *153*
Boyd-Barrett, O., 30, *153*
Brody, R. A., 31, *153*
Buddenbaum, J. M., 10, *158*
Buergenthal, T., *153*

C

Chandler, O., 95, *153*
Chandler, R., 118, *153*
Charles, J., 96, *153*
Cohen, B. C., 31, 32, 130, 134, 135, 140, 141, *153*
Cole, R. R., 36, 96, *158*
Collingwood, C., 133, *153*
Corrigan, W. T., 118, *153*
Crystal, L. M., 6, 121, *153*
Cutlip, S. M., 24, *153*

D

Davison, W. P., 31, 135, 141, *153*
Diamond, E., 136, *153*
Dizard, W., *153*

E

Edwardson, M., 11, *154*
Ehrlich, E., 118, *154*
Elliott, P., 18, 29, 36, 96, 101, *154*
Entman, R. M., 31, 32, 136, 140, *156*
Epstein, E. J., 5, 7, 9, 13, 21, 26, 27, 28, 29, 47, 102, 116, 131, *154*

F

Feders, S., 116, *154*
Fenton, T., 6, 13, 131, *154*
Fishman, M., 26, 29, *154*
Frank, R. S., 35, 40, 144, 149, *154*
Friendly, F., 7, *154*
Frost, R., 11, *158*

G

Galtung, J., 18, 21, 22, 23, 27, 136, *154*
Gandy, O. H., Jr., 140, *154*
Gans, H. J., 21, 29, 136, *154*
Gerbner, G., 51, 96, *154*
Golding, P., 18, 29, 36, 96, 101, *154*
Griggs, H., 3, 5, 6, 12, 13, 31, 44, 118, 131, 132, 133, *156*
Grooms, D., 11, *154*

H

Hardy, A., 36, *155*
Harris, P., 29, 30, 38, 52, 96, 113, 114, *154*
Hester, A., 36, 96, 97, *154*
Hickey, N., 137, *154*
Holsti, O. R., 36, *155*
Hulten, O., 4, *155*

I
Iyengar, S., 142, 143, *155*

J
Joblove, M., 131, *152*

K
Kastelnik, C., 118, *155*
Katz, E., 11, 20, *155*
Kinder, D. R., 142, 143, *155*
King, J., 1, *155*
Kliesch, R. E., *155*

L
Lang, G. E., 22, *155*
Lang, K., 22, *155*
Larson, J., 36, *155*
Larson, J. F., 35, 36, 38, 57, 69, 70, 81, 96, 116, 118, 119, 121, *155*, *156*
Lichty, L. W., 8, *156*
Lippmann, W., 22, 51, *156*

M
Mankekar, D. R., 17, *156*
Markoff, J., *156*
Marvanyi, G., 51, 96, *154*
Matta, F. R., 16, *156*
McAnany, E. G., 69, *156*
McCombs, M. E., 140, 141, 142, *156*
Merz, C., 22, *156*
Mills, C. W., 32, *156*
Mosettig, M., 3, 5, 6, 12, 13, 31, 44, 118, 131, 132, 133, *156*

N
Nordenstreng, K., 15, *156*

O
Ostgaard, E., 26, *156*

P
Paletz, D. L., 31, 32, 47, 136, 140, *156*
Parness, P., 11, *155*
Pasadeos, Y., 96, *156*
Pearce, A., 6, 7, *156*
Pearson, R., 47, *156*
Peters, M. D., 142, 143, *155*
Peterson, T., 31, *157*
Plante, J. F., 118, *157*
Pollock, J. C., 134, *157*
Powers, R., 7, *157*
Proudlove, S., 11, *154*

Q
Quint, B., 4, *157*

R
Richstad, J., 17, 19, 20, *157*
Roper, B. W., 12, *157*
Rosenblum, M., 18, 19, 94, 102, 115, *157*
Rosengren, K. E., 22, 23, 96, 117, *157*
Ruge, M. H., 18, 21, 22, 23, 27, 136, *154*
Rybolt, W., 11, *158*

S
Schiller, H. I., *157*
Schlesinger, P., 21, 23, 29, *157*
Schramm, W., 3, 31, 35, 96, *157*
Schreibman, F., *152*
Semmel, A. K., 96, *157*
Shanor, D. R., 31, *153*
Shaw, D. L., 36, *158*
Sheehan, W., 4, *157*
Shore, L., 96, *153*
Siebert, F. S., 31, *157*
Signitzer, B., *157*
Somavia, J., *157*
Stauffer, J., 11, *158*
Stevenson, R. L., 36, 96, *158*
Storey, J. D., 69, 70, 81, 136, 137, *155*, *156*, *158*

T
Todd, R., 96, *153*
Townley, R., 4, 11, 132, *158*
Tuchman, G., 23, 25, 26, 29, 30, *158*
Tunstall, J., 17, 19, *158*

V
Varis, T., 15, *156*

W
Weaver, D. H., 10, 96, 97, *158*
Weaver, P. H., 28, 47, *159*
Wedell, 20
Weisman, J., 133, *159*
Westin, A., 5, 10, 29, 31, 47, 133, *159*
White, K. P., *158*
White, T., 129, 143, *159*
Wilhoit, G. C., 96, 97, *158*
Wolzien, T., 131, *159*

Y
Yu, F. T. C., 31, *153*

Subject Index

A

ABC. *See* Network television

Africa
coverage of, 61, 88–90
compared to other regions, 63–71
nations and territories of, 173–174

Agencies, news, 16–17, 30, 146–147
AP and UPI, 115–116, 118–121,
125–128, 135, 141

Agenda-setting, 140–143

Air time, 41–42

American Newspaper Publishers
Association, 9

Amount of coverage, 40–42
changes in, 105, 106
of developed vs. developing nations,
98–100
major findings on, 145
by nation, 57–58

Analysis, units of, 37–38, 117, 166–167. *See
also* Research

Anchor reports
on ABC and CBS, 125, 126
category of nation mentioned in,
101–102
as dependent variable, 117–118,
127–128
on developed, developing and socialist
nations, 107–109
flows of, 115–116
format, 37, 42–45
interregional comparison of coverage
with, 64–67
rank ordering of, 47–49
by region, 71–73, 76–77, 82, 86–87
time devoted to, 41
trends in, 146

Arab-Israeli war, 131

Asia
coverage of, 78–81
compared to other regions, 63–71
nations and territories of, 176–177

Associated Press (AP), 115–116, 118–121,
125–128, 135, 141

Audience flow, 25, 47

B

Begin, M., 149

Bergman, J., 138

Bivariate correlations, 125

Bivariate regression analyses, 179

"Blind spots" of network television,
111

Bureau location, 114, 116, 118, 120, 121,
125–128

C

Cable News network (CNN), 10–11

Canada, coverage of, 63–71

Carter, J. E., 137

Catalyst role, 140–143, 149

Categorization of nations, 97, 119

CBS. *See* Network television

Chain of news communication, 21,
113–115

Changes in coverage over time, 91–92,
103–109

Chou En Lai, 139

CNN, 10–11

Coding content data, 167–169

Coe, D., 3

Committee on Foreign Relations, 4

Concentration in news geography, 90–91,
146

Conceptual approach to study of TV news, 20–33
   importance of context for, 32–33
   for network TV, 23–32
   technical structure, 113–121, 125
   theoretical approaches, 20–23
Content, international news, 34–50. *See also* Thematic content
   amount of, 40–42
   formats, 42–45
   methods of analysis, 34–39
   packaging of, 49–50
   rank ordering of stories, 46–49
Content analysis, 34, 166–169. *See also* Research
Content research, 20
Context of study, 32–33
Correlations, bivariate, 125
Costs, transmission, 3, 5
Creating Reality (Altheide), 23
Crisis content
   changes in, 109, 110
   in developed vs. developing nations, 102–103, 104, 112
   findings concerning, 146
   themes, 45–46, 69–70, 74–75, 78, 81
   via satellite, 131
Cronkite, W., 149

D

Developed vs. developing nations, coverage of, 93–112
   categorization, 97
   changes over time in, 103–109
   crisis coverage, 102–103, 104, 112
   findings on, 147–148
   nations mentioned, 100–101
   patterns of, 109–112
   problem of, 93–95
   quantity of, 98–100
   reporting formats, 101–102
   research on, 95–96
Development news, 18, 94, 143, 150
Diplomacy, international, 13–14, 31–32
Domestic video reports
   category of nation mentioned in, 101–102
   on developed, developing and socialist nations, 107–109
   format, 37, 42–45
   interregional comparison of coverage with, 64–67

Domestic video reports (cont.)
   rank ordering of, 47–49
   by region, 71–73, 76–77, 82, 86, 88, 147
   time devoted to, 41
   trends in, 146
Donaldson, S., 139

E

Eastern Europe
   coverage of, 61, 82–84
     compared to other regions, 63–71
     nations and territories of, 174
Economics of network television news, 6–7
Effects research, 20
Electronic newsgathering technology, 5–6, 121, 130, 132
Elites, foreign policy, 135–139
Explanatory studies, 96–97
Extramedia data, 22, 117

F

Failure, tendency to focus on, 94
Flow, news, 23–26, 95–96
Ford, G., 137
Foreign correspondents, 132–134
Foreign policy process, TV news role and, 129–143
   as catalyst, 140–143, 149
   changes in, 13–14, 31–32
   as observer, 130–134, 148
   as participant, 134–139, 148–149
Foreign video reports
   category of nation in, 101–102
   as dependent variable, 117–118
   on developed, developing and socialist nations, 107–109
   flows of, 115–116
   format, 37–38, 42–45
   interregional comparison of coverage with, 64–67
   multiple regression analysis of, 127–128
   national origin of, 58–60, 66–67, 100, 103–106, 111, 147–148
   rank ordering of, 47–49
   by region, 71–73, 76–77, 82, 84–86, 88, 147
   regional patterns, 147
   satellite communication influence on, 121–125
   time devoted to, 41
   trends in, 146

Formats, report. *See also* specific reports
  changes in, 105–109
  for coverage by region, 71–73, 75–77,
    80–82, 84–87, 88
  for developed vs. developing nation
    coverage, 101–102
  interregional comparison, 62–67
  major, 42–45
  story level analysis and, 37–38
  trends in, 146
Free flow of information doctrine, 14–16

G
Geography, news, 51–92
  major feature of, 146–147
  nations and territories on network
    news, 53–55, 101
  patterns in coverage, 90–92, 146
  regions, coverage of, 61–90
    Africa, 61, 88–90
    Asia, 78–81
    Eastern Europe and USSR, 61, 82–84
    interregional comparison, 62–71,
      146–147
    Latin America, 61, 84–87
    Middle East, 75–78
    Western Europe, 71–75
    world news leaders, 55–60, 181–187
Global impact of U.S. media, 19–20
Government control, 18–19, 132
Gross National Product Per Capital, 119,
  120, 125–128

H
Hierarchy in news geography, 90–91, 146

I
Intelsat global satellite system, 2–4, 44,
  118–128
Interactive format, 25
International Commission for the Study of
  Communication Problems, 19, 92–93
International news
  available through other media, 10–11
  definition of, 35–36
  nature of coverage, 145–149
International structures, 30
Interregional coverage comparison, 62–71,
  146–147
Intramedia data, 22
Iran Hostage Crisis, 4, 12, 13–14

J
Jarriel, T., 138
John Paul I, Pope, 74
Johnson, L. B., 135

K
Kekkonen, U., 15
Kissinger, H., 13, 137

L
Latin America
  coverage of, 61, 84–87
    compared to other regions, 63–71
  nations and territories of, 171–172
Leaders, world news, 55–60, 181–187
Local television stations, 11
Longitudinal studies, 34–35. *See also*
  Research

M
MacBride Commission, 19, 92–93
"MacNeil-Lehrer Report", 25
Magazines, news, 10
Making News (Tuchman), 23
Mao Tse-Tung, 139
Middle East
  coverage of, 75–78
    compared to other regions, 63–71
  nations and territories of, 175–176
Multiple regression analysis, 125–128

N
Nation level analysis, 38
Nations. *See also* Developed vs. developing
    nations, coverage of; Geography news
  categorization of, 97, 119
  differences in coverage of, 91
  listing of, 171–178
  mentioned in news, 53–58, 100–101,
    181–187
Network television
  ABC and CBS
    anchor reports on, 125, 126
    bureau locations, 118
    foreign video reports on, 58–60,
      66–67, 122–123, 126
    most frequently mentioned nations on,
      55–56, 182–183
  conceptual approach for, 23–32
  economics of news, 6–7
  expansion of role, 1–2

Network television (cont.)
    international concerns affecting U.S.,
        14–20
    NBC, 55–56, 186–187
    newspapers vs., 7–12, 28
News
    agencies, 114, 146–147
    factors, 21–22
    flow, 23–26, 95–96
    process, 26–27, 29
    structure, 27–30
    values, 17–18
Newscast level analysis, 38
News conference, presidential, 136–137
Newspapers, 7–12, 28
Newsweek (magazine), 145
New World Communication Order, 14–16
New York Times, 135, 141, 145
"Nightline", 25
Nixon, R., 13–14, 135, 137–139
Noncrisis themes, 45–46, 69–70
North Africa, 175–176
North America, nations and territories of,
    171

O
Observer role, 130–134, 148

P
Pacific, nations and territories of, 176–177
Packaging international news, 49–50
Participant role, 134–139, 148–149
Paul VI, Pope, 74
Political difficulties, 4
Population data, 119, 120, 125–128
President of U.S., contribution of, 136–139
Press and Foreign Policy, The (Cohen), 130
Production research, 20
Profitability of network news, 6–7
"Pseudo-events", 26
Public opinion, international news and, 31
Putting 'Reality' Together (Schlesinger), 23

Q
Quantification of findings, 39

R
Radio, 10
Rank ordering of stories, 46–49
Reagan, R., I, 37, 135–136, 137
Reasoner, H., 138, 139

Regions. *See* Geography, news; Nations
Regression analyses, 125–128, 179
Reliability of research, 36–37, 170
Report formats. *See* Formats, report
Research
    on agenda-setting, 140–143
    on developed vs. developing nation
        coverage, 95–96
    on international news, 20–23
    methods, 34–39
    on use of newspapers vs. TV, 7–12
Roper Organization, 8, 12
Roper Survey, 8–10

S
Sadat, A., 13, 149
Sampling, 35, 165
Satellites
    crisis reporting via, 131
    earth stations, 44, 114, 118–121,
        125–128
    influence of, 2–5, 121–125
Saturation coverage, 12, 50
Sirica, Judge, 44
Socialist nations, coverage of, 99–109,
    147–148
Soviet Union. *See* USSR
Stewart, B., 133–134
Story(ies)
    format. *See* Formats, report
    level analysis, 37–38
    rank ordering of, 46–49
Structure, news, 27–30
Studies, *See* Research

T
Technical structure, 113–128, 148
    conceptual approach to, 113–121, 125
    multiple regression analysis of,
        125–128
Technological developments, 2–6
    electronic newsgathering, 5–6, 121,
        130, 132
    foreign policy news and, 130–134, 148
    governmental control and, 132
    satellites, 2–5, 44, 114, 118–128, 131
Television Information Office, 8, 12
Television News Index and Abstracts, 34,
    36–37, 137, 166, 168, 170
Territories, 53–55, 171–178. *See also*
    Geography, news; Nations

Thematic content, 45–46
  interregional comparison of, 69–71
  in regional coverage, 74–75, 78, 81, 84, 87, 90
Theoretical approaches to international news, 20, 23
Third World coverage. *See* Developed vs. developing nations, coverage of
Third World News Agency, 17
Time
  changes in coverage over, 91–92, 103–109
  devoted to international news, 41–42
  dimension of research, 34–35
  value of news, 11
Time (magazine), 145
Transmission costs, 3, 5
Transnational structures, 30
Travel, presidential, 136, 137–139
Trends
  findings on, 145–149
  in regional coverage, 67–69
    African, 88–90
    Asian, 81
    Eastern Europe and USSR, 82–84
    Latin American, 87
    Middle East, 77–78
    Western Europe, 73–74

U
UNESCO, 3, 15, 19
United Press International (UPI), 115–116, 118–121, 125–128, 135, 141
United States, reference to, 53–55
Units of analysis, 37–38, 117, 166–167
Universal Declaration of Human Rights, 15
UPI-TN, 17
USSR, 60, 82–84, 174

V
Values, news, 17–18
Vanderbilt Television News Archive, 142, 144
Videotape, use of, 5–6, 44
Vietnam war, 69, 70, 80–81, 131
Visnews, 17

W
War coverage, 13, 131. *See also* specific wars
Washington Post, The, 135, 141
Watergate scandal, 44
Western Europe
  coverage of, 71–75
    compared to other regions, 63–71
    nations and territories of, 174–175
"World: The Clouded Window" (program), 51–52
World news leaders, 55–60, 181–187